John Wesley, O. W. Gordon

Rev. John Wesley's Valuable Primitive Remedies

An easy and natural method of curing most diseases

John Wesley, O. W. Gordon

Rev. John Wesley's Valuable Primitive Remedies
An easy and natural method of curing most diseases

ISBN/EAN: 9783337390006

Printed in Europe, USA, Canada, Australia, Japan

Cover: Foto ©berggeist007 / pixelio.de

More available books at **www.hansebooks.com**

REV. JOHN WESLEY'S

VALUABLE

PRIMITIVE REMEDIES,

OR AN

EASY AND NATURAL METHOD OF CURING MOST DISEASES.

ALSO

MODERN MEDICINE,

BY

O. W. Gordon, M. D.,

WITH USEFUL AND VALUABLE RECEIPTS.

For Books, address

O. W. GORDON, Editor and Manager.

CHICAGO,

1880.

Entered according to Act of Congress, in the year 1879, by
S. BEACH, (Sarah Beach.)
In the office of the Librarian of Congress, at Washington. D. C.

WILLIAM H. WILSON, PRINTER,
ROOM 26, 182 S. CLARK ST.,
CHICAGO, ILL.

CHICAGO STEREOTYPE WORKS.
85 & 87 FIFTH AVENUE,
CHICAGO, ILLS.

TO

The Struggling Millions

ON WHOSE SHOULDERS THE BURDENS AND HOPES OF
AMERICAN INSTITUTIONS REST,

THIS VOLUME IS RESPECTFULLY DEDICATED.

AUTHOR'S PREFACE.

After more than one critical examination of Rev. John Wesley's Primitive Medicine, the writer pronounces it a master-piece of simplicity and truthfulness. Entirely worthy of its eminent author. Stripped of everything save unvarnished facts.

Wesley was too far seeing and clear in his perceptions of truth, to allow the "so-called philosophy" of the day to mar his work of love in doing good to the *bodies* as well as the souls of mankind.

Some of his statements seem almost prophetic. On some points he was more than a century in advance of his times.

Not a single page of his preface should be left unread, it will richly repay for the time consumed in its careful perusal, and a hundred times repay if the advice therein given is put to practical use by the reader.

While the author claims many portions of the second part of this book as original, yet he freely acknowledges himself under obligation to a great number of medical writers of every school of medicine, for many ideas of priceless value to be found in this volume. He has endeavored to put these gathered truths in a correlative form.

As there is unity and harmony in all the works of creation ; (as far as we have any positive knowledge) therefore, in studying man in the light of medical science, we must study from the broad plain of harmonious unity, as shown in other parts of the visible creation.

Let *known facts* govern our theory and practice; let no theory distort the truth; then shall we deal safely with the physical well-being of man; both in sickness and health.

As in learning in general, true medical knowledge does not consist in the *amount* of information possessed, but in its truthfulness: for *truth* is the embodiment of wise-knowledge.

In the treatment as given for the various diseases enumerated in this book; the author has endeavored to be

1st. Plain.
2d. Concise.
3d. Practical.
4th. Reliable.

Nothing has been rejected *because* it was old, nor accepted *because* it was new, but all gathered from both the old and the new.

The study has been to make the book a *household treasure*, so that when *once used* in the home, it will be *consulted* again, and again.

While the author has endeavored to give the readers of this book the very best remedies and treatment as known at the present day, every dangerous remedy or treatment has been carefully excluded from its pages.

A repetition in the second of what has been recomended in the first part of the work, has been avoided as far as practical, by simply referring to the page and section where it will be found.

While endorsing and recomending the general practice of the hygienic rules given by Mr. Wesley for the help of the sick as well as the prevention of disease: It is a privilege as well as duty to give a plain statement of a few of the principal, general laws which govern the action of the human organism, both in health and disease; that in the light of these known principles, any person of ordinary judgment can intelligibly treat most diseases with confidence and success: (but in complicated cases; advise that a competent Christian physician be called.)

Hear the conncil of Salerno's school (in the ninth century.)

> "If thou to health and vigor would attain,
> Shun weighty cares—all anger deem profane,
> From heavy suppers and from wine abstain,
> Nor trivial count it, after pompous fare,
> To rise from table and to take the air.
> Shun idle, noonday slumber, nor delay
> The urgent calls of nature to obey.
> These rules if thou wilt follow to the end,
> Thy life to greater length mayst extend.
> Shouldst thou doctors need; these in doctors stead—
> Rest, cheerfulness, and table thinly·spread."

How concise and true? and this from the great medical college of the ninth century, whose professorships were shared by women whose wisdom has hardly been surpased even in the nineteenth century.

While there has been much done to put a knowledge of the laws of health and the careful care of the sick into the hands of the people: yet there has been a lack of adaptability to the needs of the people, the writers are generally too diffuse, redundant and verbose for the masses.

While there are noble exceptions, many in the profession seem to fear that it will lower the dignity of their calling and be the means of a greater number of women entering on the study and practice of medicine, to the great injury of masculine dignity and profit. So let it be! For no greater blessing in this direction could befall us as a civilized nation, than that every town and village in this "land of the free and home of the brave" should have the professional services of a competent woman physician.

Says one of our best medical writers, and a professor in a medical college: "One thing is remarkable in reference to safe medication,—that the people are in advance of the profession. They seem to be less heroic in the use of poisons than the latter, and when they prescribe for themselves they usually employ very simple and harmless agents, and hence occasion much less mischief than might be apprehended. They have an opportunity also of observing the peculiarities of their own habits, susceptibilities and necessities; and when informed on the subject of medicine as much as they are on other subjects generally, they will in many instances prescribe more successfully for themselves than physicians who have less chance to know their cases."

> "Good common sense with quiet care cures disease,
> Not empty words of boastful lying quack;
> The first combined gives suff'ring mortals ease.
> The last to perish, leaves them on the rack."—*Salernum*.

The author has written what he considered to be a duty and privilege, and leaves the results with a discerning public and All Ruling Providence.

The appendix will be found to contain valuable receipts and other matters of general interest, which no family can afford to dispense with for many times the price of the book.

O. W. GORDON, M. D.

PREFACE.

WHEN man came first out of the hands of the Great Creator, clothed in body, as well as in soul, with immortality and incorruption, there was no place for physic, or the art of healing. As he knew no sin, so he knew no pain, no sickness, weakness, or bodily disorder. The habitation wherein the angelic mind, the Divine Particulæ Auræ, abode, although originally formed of the dust of the earth, was liable to no decay. It had no seeds of corruption or dissolution within itself; and there was nothing without to injure it; heaven and earth, and all the host of them were mild, benign, and friendly to human nature. The entire creation was at peace with man, so long as man was at peace with his Creator. So that well might the morning stars sing together, and all the sons of God shout for joy.

2. But since man rebelled against the Sovereign of heaven and earth, how entirely is the scene changed? The incorruptible frame hath put on corruption, the immortal hath put on mortality. The seeds of wickedness and pain, of sickness and death, are now lodged in our inmost substance; whence a thousand disorders continually spring, even without the aid of external violence. And how is the number of these increased by every thing round about us? The heavens, the earth, and all things contained therein, conspire to punish the rebels

against their Creator. The sun and moon shed unwholsome influences from above; the earth exhales poisonous damps from beneath; the beasts of the field, the birds of the air, the fishes of the sea, are in a state of hostility; the air itself that surrounds us on every side, is replete with shafts of death; yea, the food we eat daily saps the foundation of that life which cannot be sustained without it. So has the Lord of All secured the execution of His decree—"Dust thou art, and unto dust shalt thou return."

3. But can nothing be found to lessen those inconveniences which cannot be wholly removed? To soften the evils of life, and prevent in part the sickness and pain to which we are continually exposed? Without question there may One grand preventative of pain and sickness of various kinds, seems intimated by the grand Author of Nature in the very sentence that entails death upon us,—" In the sweat of thy face shalt thou eat bread, till thou return to the ground. The power of exercise, both to preserve and restore health, is greater than can well be conceived; especially in those who add temperance thereto, who, if they do not confine themselves altogether to eat either " Bread or the herb of the field," (which God does not require them to do) yet steadily observe both that kind and measure of food which experience shows to be most friendly to health and strength.

4. It is probable Physic, as well as Religion, was in the first ages chiefly traditional; every father delivering down to his sons what he had in like manner received, concerning the manner of healing both outward hurts and the diseases incident to each climate, and the medicines which were of the

greatest efficacy for the cure of each disorder. It is certain this is the method wherein the art of healing is preserved among the American Indians to this day. Their diseases indeed are exceeding few; nor do they often occur, by reason of their continual exercise, and (till of late) universal temperance. But if any are sick, or bit by a serpent, or torn by a wild beast, the fathers immediately tell their children what remedy to apply. And it is rare that the patient suffers long; those medicines being quick, as well as generally infallible·

5. Hence it was, perhaps, that the ancients, not only of Greece and Rome, but even of barbarous nations, usually assigned physic a divine original. And indeed it was a natural thought, that He who had taught it to the very beasts and birds, the Cretan Stag, the Egyptian Ibis, could not be wanting to teach man.

Sanctius his Animal, *mentisque capacius altæ*.

Yea, sometimes even by those meaner creatures, for it is easy to infer, "If this will heal that creature, whose flesh is nearly of the same texture with mine, then in a parallel case it will ·heal me." The trial was made—the cure was wrought—and experience and physic grew up together.

6. And has not the Author of Nature taught us the use of many other medicines by what is vulgarly termed accident?' Thus, one walking some years since in a grove of pines, at a time when many in the neighboring towns were afflicted with a kind of new distemper—little sores in the inside of the mouth —a drop of the natural gum fell from one of the trees on a book which he was reading. This he took up, and thoughtlessly applied to one of those sore places. Finding the pain immediately cease, he applied it to another, which was also

presently healed. The same remedy he afterwards imparted to others, and it did not fail to heal any that applied it. And doubtless numberless remedies have been thus casually discovered in every age and nation.

7. Thus far physic was wholly founded on experiment. The European, as well as the American said to his neighbor, "Are you sick? Drink the juice of this herb and your sickness will be at an end. Are you in a burning heat? Leap into that river and then sweat till you are well. Has the snake bitten you? Chew and apply that root, and the poison will not hurt you." Thus, ancient men, having a little experience joined with eommon sense and common humanity, cured both themselves and neighbors of most of the distempers to which every nation was subject.

8. But in process of time, men of a philosophical turn were not satisfied with this. They began to enquire how they might account for these things? How such medicines wrought such effects? They examined the human body and all its parts; the nature of the flesh, veins, arteries, nerves; the structure of the brain, heart, lungs, stomach, bowels; with the springs of the several kinds of animal functions. They explored the several kinds of animal and mineral, as well as vegetable substances; and hence the whole order of physic, which had obtained to that time, became inverted. Men of learning began to set experience aside—to build physic upon hypothesis—to form theories of diseases and their cure, and to substitute these in the place of experiments.

9. As theories increased, simple medicines were more and more disregarded and disused. till in a course of years th

greater part of them were forgotten, at least in the politer nations. In the room of these, abundance of new ones were introduced, by reasoning, speculative men ; and those more and more difficult to be applied, as being more remote from common observation. Hence, rules for the application of these, and medical books, were immensely multiplied, till at length physic became an abtruse science, quite out of the reach of ordinary men.

10. Physicians now began to be in admiration, as persons who were something more than human. And profit attended employ as well as honor ; so that they had now two weighty reasons for their keeping the bulk of mankind at a distance, that they might not pry into the mysteries of the profession. To this end, they increased those difficulties by design, which began in a manner by accident. They filled their writings with abundance of technical terms, utterly unintelligible to plain men. They affected to deliver their rules, and to reason upon them in an abstruse and philosophical manner. They represented the critical knowledge of Astronomy, Natural Philosophy, (and what not ?) Some of them insisting upon that of Astronomy, and Astrology too, as necessary previous to the understanding of the art of healing. Those who understood only how to restore the sick to health, they branded with the name of Empirics. They introduced into practice abundance of compound medicines consisting of so many ingredients, that it was scarce possible for common people to know which it was that wrought the cure ; abundance of exotics; neither the nature nor names of which their own countrymen understood ; of chemicals, such as they neither had skill, nor fortune, nor time to prepare ; yea, and of

dangerous ones, such as they could not use without hazarding life, but by the advice of a physician. And thus both their honor and gain were secured, a vast majority of mankind being utterly cut off from helping either themselves or their neighbors, or once daring to attempt it.

11. Yet there have not been wanting from time to time, some lovers of mankind, who have endeavored, even contrary to their own interest, to reduce physic to its ancient standard; who have labored to explode it out of all the hypothesis and fine-spun theories, and to make it plain intelligible thing, as it was in the beginning; having no more mystery in it than this—" Such a medicine removes such a pain." These have demonstrably shewn, that neither the knowledge of Astrology, Astronomy, Natural Philosophy, nor even Anatomy itself is absolutely necessary to the quick and effectual cure of most diseases incident to human bodies; nor yet any chemical, or exotic, or compound medicine, but a simple plant or root duly applied. So that every man of common sense, unless in some rare case, may prescribe either to himself or neighbor; and may be very secure from doing harm, even where he can do no good.

12. Even in the last age there was something of this kind done, particularly by the great and good Dr. Sydenham; and in the present, by his pupil, Dr. Dover, who has pointed out simple medicines for many diseases. And some such may be found in the writings of the learned and ingenious Dr. Cheyne; who doubtless would have communicated more to the world, but for the melancholy reason he gave one of his friends, that pressed him with some passages in his works which too much countenanced the modern practice, " O Sir,

we must do something to oblige the faculty, or they will tear us in pieces."

13. Without any regard to this, without any concern about the obliging or disobliging any man living, a mean hand has made here some little attempt towards a plain and easy way of curing most diseases. I have only consulted herein, experience, common sense, and the common interest of mankind. And supposing they can be cured this easy way, who would desire to use any other? Who would not wish to have a physician always in his house, and one that attends without fee or reward? To be able (unless in some few complicated cases) to prescribe to his family as well as himself.

14. If it be said, but what need is there of such attempt? I answer the greatest that can possibly be conceived. Is it not needful, in the highest degree, to rescue men from the jaws of destruction? From wasting their fortunes, as thousands have done, and continue to do daily? From pining away in sickness and pain, either through the ignorance or dishonor of physicians. Yea, and many times throw away their lives after their health, time, and substance.

Is it enquired, but are there not books enough already on every part of the art of medicine? Yes, too many ten times over, considering how little to the purpose the far greater part of them speak. But besides this, they are too dear for poor men to buy, and too hard for plain men to understand.

Do you say, "But there are enough of those collections of receipts." Where? I have not seen one yet, either in our own or any other tongue, which contains only safe, and cheap, and easy medicines. In all that have yet fallen into my hands, I find many dear and many far-fetched medicines;

besides many of so dangerous a kind as a prudent man would never meddle with. And against the greater part of these medicines there is a further objection—they consist of too many ingredients. The common method of compounding or re-compounding medicines can never be reconciled to common sense. Experience shews, that one thing will cure most disorders, at least as well as twenty put together. Then why do you add nineteen? Only to swell the apothecary's bill. Nay, possibly, on purpose to prolong the distemper, that the doctor and he may divide the spoil.

But admitting there is some quality in the medicine proposed which has need to be corrected, will not one thing correct it as well as twenty? It is possible, much better And if not, there is a sufficiency of other medicines which need no such correction.

How often, by thus compounding medicines of opposite qualities, is the virtue of both utterly destroyed? Nay, how often do those joined together destroy life, which singly, might have preserved it? This occasioned that caution of the great Boerhaave, against mixing things without evident necessity, and without full proof of the effect they will produce when joined together, as well as of that they produce when asunder; seeing (as he observes) several things which, separately taken, are safe and powerful medicines, when compounded, not only lose their former powers, but commence a strong and deadly poison.

PRIMITIVE REMEDIES.

MANNER OF USING THE MEDICINES.

As to the manner of using the medicines here set down, I would advise, as soon as you know your distemper, which is very easy unless in a complication of disorders, and then you would do well to apply to a physician that fears God. For one that does not, be his fame ever so great, I should expect a curse rather than a blessing.

First.—Use the first of the remedies for that disease which occurs in the ensuing collection, unless some other of them be easier to be had, and then it may do just as well.

Secondly.—After a competent time, if it takes no effect, us the second, and the third, and so on. I have purposely set down, in most cases, several remedies for each disorder, not only because all are not equally easy to be procured at all times, and in all places, but likewise, because the medicine which cures one man will not always cure another of the same distemper. Nor will it cure the same man at all times. Therefore it was necessary to have a variety. However, I have subjoined the letter (I) to those medicines which some think are infalliable. (Tried) to those which I have found to be of the greatest efficacy. I believe many others to be of equal virtue, but it has not laid in my way to make the trial. One I must aver from personal knowledge, grounded on a thousand experiments, to be far superior to all other medicines I have known—I mean electricity. I cannot but entreat those who are well-wishers to mankind to make full proof of this. Certainly it comes the nearest to a universal medicine of any yet known in the world. Most of the medicines which I prefer to the rest are now marked with an asterisk.*

Thirdly.—Observe all the time the greatest exactness in your regimen or manner of living. Abstain from all mixed or high-seasoned food. Use plain diet easy of digestion, and this as sparingly as you can consistently with ease and strength.—Drink only water if it agrees with your stomach. Use as much exercise daily in the open air as you can, without weariness. Sup at six or seven on the lightest food; go to bed early, and rise betimes. To persevere with steadiness in this course, is often more than half the cure. Above all, add to the rest, for it is not labor lost, that old-fashioned medicine—*prayer;* and have faith in God, who " Killeth and maketh alive, who bringeth down to the grave and bringeth up."

For the sake of those who desire, through the blessing of God, to retain the health which they have recovered, I have added a few plain easy rules, briefly transcribed from Dr. Cheyne.

PLAIN EASY RULES.

1. (I) The air we breathe is of great consequence to our health. Those who have been long abroad in easterly or northerly winds should drink some warm pepper tea on going to bed, or a draught of toast and water.

2. Tender people should have those who lie with them, or are much about them, sound, sweet, and healthy.

3. Every one that would preserve health should be as clean and sweet as possible in their houses, clothes, and furniture.

II. 1. The great rule of eating and drinking is to suit the quality and quantity of the food to the strength of the digestion; to take always such a sort and such a measure of food as sits light and easy on the stomach.

2. All pickled, or smoked, or salted food, and all high-seasoned, is unwholesome.

3. Nothing conduces more to health than abstinence and plain food, with due labor.

4. For studious persons, about eight ounces of animal food, and twelve of vegetable, in twenty-four hours, is sufficient.

5. Water is the wholesomest of all drinks; it quickens the the appetite and strengthens the digestion most.

6. Strong, and more especially spiritous liquors, are a certain, though slow, poison.

7. Experience shows there is very seldom any danger in leaving them off all at once.

8. Strong liquors do not prevent the mischiefs of a surfeit, or carry it off so safely as water.

9. Malt liquors are extremely hurtful to tender persons.

10. Coffee and tea are extremely hurtful to persons who have weak nerves.

III. 1. Tender persons should eat very light suppers, and that two or three hours before going to bed.

2. They ought constantly to go to bed about nine, and rise at four or five.

IV. 1. A due degree of exercise is indispensably necessary to health and long life.

2 Walking is the best exercise for those who are able to bear it; riding for those who are not. The open air, when the weather is fair, contributes much to the benefit of exercise.

3. We may strengthen any weak part of the body by constant exercise. Thus the lungs may be strengthened by loud speaking, or walking up an easy ascent; the digestion and the nerves by riding; the arms and hams by strong rubbing them daily.

4. The studious ought to have stated times for exercise, at least two or three hours a day; the one-half of this before dinner, the other before going to bed.

5. They should frequently shave, and frequently wash their feet.

6. Those who read or write much, should learn to do it standing; otherwise it will impair their health.

7. The fewer clothes any one uses by day or night, the hardier he will be.

8. *Exercise, first, should be always on an empty stomach; secondly, should never be continued to weariness; thirdly, after it, we should take to cool by degrees, otherwise we shall catch cold.

9. The flesh brush is a most useful exercise, especially to strengthen any part that is weak.

10. Cold bathing is of great advantage to health; it prevents abundance of diseases. It promotes perspiration, helps the circulation of the blood, and prevents the danger of catching cold. Tender persons should pour water upon the head before they go in, and walk swiftly. To jump in with the head foremost is too great a shock to nature.

V. 1. Costiveness cannot long consist with health; therefore care should be taken to remove it at the beginning, and, when it is removed, to prevent its return by soft, cool, opening diet.

2. Obstructed perspirations (vulgarly called catching cold) is one great source of diseases. Whenever there appears the least sign of this, let it be removed by gentle sweats.

VI. 1. The passions have a greater influence upon health than most people are aware of.

2. All violent and sudden passions dispose to, or actually throw people into acute diseases.

3. Till the passion which caused the disease is calmed, medicine is applied in vain.

4. The love of God, as it is the sovereign remedy of all miseries, so in particular it effectually prevents all the bodily

---

* In malarial districts, always take a cup of coffee as soon as possible after rising from bed in the morning.

disorders the passions introduce, by keeping the passions themselves within due bounds ; and by the unspeakable joy and perfect calm serenity and tranquility it gives the mind, it becomes the most powerful of all the means of health and long life.

AN EASY AND NATURAL METHOD

OF CURING MOST DISEASES.

---

1. ABORTION, (TO PREVENT.)

Women of a weak or relaxed habit should use solid food, avoiding great quantities of tea and other weak and watery liquors. They should go soon to bed and rise early, and take frequent exercise but avoid being over-fatigued.

If of full habit, they ought to use a spare diet, and chiefly of the vegetable kind, avoiding strong liquors and everything that may tend to heat the body, or increase the quantity of blood.

In the first case, take daily half pint of decoction of Lignum Guaiacum; boiling an ounce of it in a quart of water for five minutes.

In the latter case, give half a drachm of powdered nitre in a cup of water-gruel, every five or six hours; in both cases she should sleep on a hard mattrass with her head low, and be kept cool and quiet.

2. FOR AN AGUE.*

Go into the cold bath just before the cold fit.

NOTHING tends more to prolong an ague than indulging a lazy, indolent disposition. The patient ought therefore, between the fits, to take as much exercise as he can bear, and to use a light diet, and, for common drink, lemonade is the most proper.

When all other means fail, give blue vitrol, from one grain to two grains in the absence of the fit, and repeat it three or four times in twenty-four hours.

Or, take a handful of groundsel, shred it small, put it into a paper bag four inches square, pricking that side which is to be next the skin full of holes ; cover this with a thin linen, and wear it on the pit of the stomach, renewing it two hours before the fit. (Tried.)

Or, apply to the stomach a large onion, slit.

Or, melt two pennyworth of frankincense, spread it on linen, grate a nutmeg upon it, cover it with linen, and hang this bag on the pit of the stomach. I have never yet known it fail.

Or, boil yarrow in new milk till it is tender enough to spread as a plaster. An hour before the cold fit apply this to the wrists, and let it be on till the hot fit is over. If another fit comes use a fresh plaster. This often cures a Quartan.

Or, drink a quart of cold water just before the cold fit, then go to bed and sweat.

---

* An ague is an intermitting fever, each fit of which is preceded by a cold shivering, and goes off in a sweat.

PRIMITIVE REMEDIES.

Or, make six middling pills of cobwebs. Take one a little before the cold fit; two a little before the next fit, (suppose the next day;) the other three, if need be, a little before the third fit. This seldom fails.

Or, put a teaspoonful of salt of tartar into a large glass of spring water, and drink it by little and little; repeat the same dose the next two days before the time of the fit.

Or, two small teaspoonful of sal prunella an hour before the fit. It commonly cures in thrice taking.

Or, a large spoonful of powdered camomile flowers.

Or, a tea-spoonful of spirits of hartshorn in a glass of water.

Or, eat a small lemon, rind and all.

In the hot fit, if violent, take eight or ten drops of laudanum; if costive, in hiera picra. (A preparation of aloes.)

Doctor Lind says, an ague is certainly cured by taking from ten to twenty drops of laudnum, with two drachms of syrup of poppies, in any warm liquid, half an hour after the heat begins.

IT IS proper to take a gentle vomit, and sometimes a purge, before you use any of these medicines. If a vomit is taken two hours before the fit is expected, it generally prevents that fit, and sometimes cures an ague, especially in children. It is also proper to repeat the medicine (whatever it be) about a week after, in order to prevent a relapse. Do not take any purge soon after. The daily use of the flesh brush, and frequent cold bathing, are of great use to prevent relapses.

Children have been frequently cured by wearing a waistcoat in which bark was quilted.*

---

* Peruvian Bark.

3. TERTIAN AGUE.*

Is often cured by taking a purge one day, and the next bleeding in the beginning of the fit.

Or, take a tea-spoonful of salt of tartar in spring water. This often cures double Tertians, tripple Quartans, and long lasting fevers, especially if senna be premised twice or thrice

Or, apply to each wrist a plaster of molases and soot. (Tried.)

Or, use the cold bath, unless of an advanced age or extremely weak. But when you use this on any account whatever, it is proper.

To bleed or purge before you begin.†

To go in cool; to emerge at once but not head foremost; to stay in only two or three minutes, or less at first.

Never to bathe on a full stomach.

To bathe twice or thrice a week at least, till you have bathed nine or ten times.

To sweat immediately after it (going to bed) in palsies, rickets, and in all diseases wherein the nerves are obstructed.

You may use yourself to it without any danger, by beginning in May, and at first just plunging in, and coming out immediately; but many have begun in winter without any inconvenience.

4. A DOUBLE TERTIAN.

Take before the fit, (after a purge or two) three ounces of cichory water, half a drachm of salt of tartar, and fifteen drops of spirit of sulphur.

---

*That is an ague which returns every other day. †Bleeding seldom used.

To perfect the cure, on the fourth day after you miss the fit, take two drachms of senna, half a drachm of salt of tartar infused all night in four ounces of cichory water. Strain it and drink it.

5 A QUARTAN AGUE.*

Apply to the suture of the head, when the fit is coming, wall July flowers, beating together leaves and flowers with a little salt. Keep it on till the hot fit is over. Repeat this if need be.

Use strong exercise (as riding or walking, as far as you can bear it) an hour or two before the fit. If possible, continue it till the fit begins. This alone will frequently cure. (Tried.)

Or, apply to the wrists a plaster of turpentine; or of bruised pepper mixed with molasses.

Or, apply oil of turpentine to the small of the back before the fit.

For a tertian or quartan, vomit with ten grains of ipecacuanha an hour before the cold fit begins; then go to bed and continue a large sweat by lemonade, (that is, lemon, sugar, and water) for six or eight hours. This usually cures in three or four times. If it does not, use the cold bath between the fits.

Or, take twenty grains of powdered saffron before the fit, in a glass of white wine.

6. ST. ANTHONY'S FIRE.†

Take a glass of tar-water warm in bed every hour, washing the part with the same.

---

* That is, an ague which misses two days; coming on Monday (suppose) and again on Thursday.

TAR-WATER is made thus:—Put a gallon of cold water to a quart of Norway tar, stir them together with a flat stick for five or six minutes. After it has stood covered for three days pour off the water clear, bottle and cork it.

Or, take a decoction of elder leaves as a sweat; applying to the part a cloth dipped in lime-water, mixed with a little camphorated spirit of wine.

LIME-WATER is made thus:—"Infuse a pound of good quick-lime in six quarts of spring water for twenty-four hours. Decant and keep it for use."

Or, take two or three gentle purges. No acute fever bears repeated purges better than this, especially when it affects the head; in the meantime boil an handful of sago, two handful of elder leaves or bark, and an ounce of alum in two quarts of forge water to a pint; wash with this every night.

If the pulse be low, and the spirits sunk, nourishing broths and a little negus may be given to advantage.

Or, let three drachms of nitre be dissolved in as much elder-flower tea as the patient can drink in twenty-four hours. If the disease attacks the head, bleeding is necessary.

Dressing the inflammation with greasy ointment, or salves etc., is very improper.

Bathing the feet and legs in warm water is serviceable, and often relieves the patient much. In Scotland the common people cover the part with a linen cloth covered with meal.

---

† St. Anthony's fire is a fever attended with red and painful swelling, full of pimples, which afterwards turn into small blisters on the face or some other part of the body. The sooner the eruption is the less danger. Let your diet be only water-gruel, or barley broth, with roasted apples.

7. THE APOPLEXY.*

To prevent, use the cold bath, and drink only water.

In the fit, put a handful of salt into a pint of cold water, and, if possible, pour it down the throat of the patient. He will quickly come to himself; so will one who seems dead by a fall. But send for a good physician immediately.

If the fit be soon after a meal, do not bleed, but vomit.

Rub the head, feet, and hands strongly, and let two strong men carry the patient upright, backward and forward about the room.

A seton in the neck, with low diet, has often prevented a relapse.

There is a wide difference between the Sanguineous and Serous Apoplexy. The latter is often followed by a palsy; the former is distinguished by the countenance appearing florid, the face swelled or puffed, and the blood vessels, especially about the neck and temples, are turgid; the pulse beats strong; the eyes are prominent and fixed; the breathing is difficult and performed with a snorting. This invades more suddenly than the Serous Apoplexy. Use large bleeding from the arm or neck, bathe the feet in warm water; cupping on the back of the head with deep scarification. The garter should be tied very tight to lessen the motion of the blood from the lower extremities.

A scruple of nitre may be given in water every three or four hours.

When the patient is so far recovered as to be able to swallow, let him take a strong purge; but if this cannot be ef-

---

* An apoplexy is a total loss of all sense and voluntary motion, commonly attended with a strong pulse, hard breathing, and snorting.

fected, a glyster should be thrown up, with plenty of fresh butter and a large spoonful of common salt in it.

In the Serous Apoplexy, the pulse is not so strong, the countenance is less florid, and not attended with so great a difficulty of breathing. Here bleeding is not necessary, but a vomit of three grains of emetic tartar may be given, and afterwards a purge as before, and the powder of white hellebore blown up the nose, etc.

This Apoplexy is generally preceded by an unusual heaviness, giddiness, and drowsiness.

8 CANINE APPETITE.*

"If it be without vomiting, is often cured by a small bit of bread dipped in wine, and applied to the nostrils."—*Dr. Scomberg.*

9. AN ASTHMA †

Take a pint of cold water every morning, washing the head therein immediately after, and using the cold bath once a fortnight.

Or, cut an ounce of stick liquorice into slices, steep this in a quart of water four and twenty hours and use it when you are worse than usual, as a common drink. I have known this give much ease.

Or, half a pint of tar-water twice a day.

Or, live a fortnight on boiled carrots only. It seldom fails.

---

* An insatiable desire of eating.

† An asthma is a difficulty of breathing from a disorder in the lungs In the common or moist asthma, the patient spits much.

*Or, take some quicksilver every morning, and a spoonful of aqua sulphurata, or fifteen drops of elixir of vitriol, in a large glass of spring water, at five in the evening. This has cured an inveterate asthma.

Or, take from ten to thirty drops of elixir of vitriol, in a large glass of spring water, three or four times a day.

ELIXIR OF VITRIOL is made thus: Drop gradually four ounces of strong oil of vitriol into a pint of spirits of wine or brandy; let it stand three days, and add to it ginger, sliced, half an ounce; Jamaica pepper, whole, one ounce. In three days more it is fit for use. But if the patient be subject to sour belchings, take the mixture for the asthmatic cough, (as Art. 57) after the elixir of vitriol.

Or, into a quart of boiling water put a tea-spoonful of balsamic æther: receive the steam into the lungs, through a fumigator, twice a day.

BALSAMIC ÆTHER is made thus: Put four ounces of spirits of wine, and one ounce of balsam of tulo, into a phial, with one ounce of æther. Keep it well corked: it will not keep above a week.

For present relief, vomit with a quart or more of warm water. The more you drink of it the better.

DO THIS whenever you find any motion to vomit; and take care always to keep your body open.

10. A DRY OR CONVULSIVE ASTHMA.

Juice of radishes relieve much; so does a cup of strong coffee, or garlic either raw or preserved, or in syrup.

---

* Only as a last resort.

Or, drink a pint of new milk morning and evening. This has cured an inveterate asthma.

Or, beat fine saffron small, and take eight or ten grains every night. (Tried.)

Take from three to five grains of ipecacuanha every morning, or five to ten grains every other evening. Do this, if need be, for a month or six weeks. Five grains usually vomit. In a violent fit, take a scruple instantly.

In an asthma, the best drink is apple water; that is, boiling water poured on sliced apples.

The food should be light, and easy of digestion. Ripe fruits baked, boiled, or roasted, are very proper; but strong liquors of all kinds, especially beer or ale, are hurtful. If any supper is taken it should be very light.

All disorders of the breast are much relieved by keeping the feet warm, and promoting perspiration. Exercise is also of very great importance, so that the patient should take as much every day as his strength will bear. Issues are found in general to be of great service.

Dr. Smith, in his Formulæ, recommends mustard whey as common drink in the moist asthma; and a decoction of madder root to promote spitting.

THE DECOCTION is made thus: Boil an ounce of madder, and two drachms of mace, in three pints of water to two pints, then strain it, and take a tea-cupful three or four times a day. But the most efficacious medicine is the quicksilver and aqua sulphurata, (as Art. 45.)

N. B. Where the latter cannot be got, ten drops of oil of vitriol in a large glass of spring water will answer the same end. I have known many persons relieved, and some cured by taking as much jalap every morning as would lie on a sixpence.

11. TO CURE BALDNESS.

Rub the part morning and evening, with onions, till it is red, and rub it afterwards with honey.

Or, wash it with a decoction of boxwood. (Tried).

Or, electrify it daily.

12. BLEEDING OF THE NOSE. (TO PREVENT.)

Drink whey largely, every morning, and eat a quantity of raisins.

Or, dissolve two scruples of nitre in half a pint of water, and take a tea-cupful every hour.

To cure it, apply to the neck behind, and on each side, a cloth dipped in cold water.

Or, put the legs and arms in cold water.

Or, wash the temples, nose, and neck, with vinegar.

Or, keep a little roll of white paper under the tongue.

Or, snuff up vinegar and water.

Or, foment the legs and arms with it.

Or, steep a linen rag in sharp vinegar, burn and blow it up the nose with a quill.

Or, apply tenets made of soft linen dipped in cold water strongly impregnated with tincture of iron, and introduced within the nostrils quite through to their posterior apertures. This method, Mr. Hay says, never failed him.

Or, dissolve an ounce of alum powdered in a pint of vinegar; apply a cloth dipped in this to the temples, steeping the feet in warm water.

In a violent case go into a pond or river. (Tried.)

13 BLEEDING OF A WOUND.

Make two or three tight ligatures toward the lower part of each joint: slaken them gradually.

Or, apply tops of nettles bruised.

Or, strew on it the ashes of a linen rag dipped in sharp vinegar and burned.

Or, take ripe puff balls, break them warily and save the powder: strew this on the wound and bind it on. (I) This will stop the bleeding of an amputated limb without any cautery.

Or, take of brandy two ounces; Castile soap, two drachms; potash, one drachm; scrape the soap fine and dissolve it in the brandy, then add the potash: mix them well together and keep them close stopped in a phial. Apply a little of this warmed to a bleeding vessel, and the blood immediately congeals.

14. SPITTING OF BLOOD.

Take a tea-cupful of stewed prunes at lying down for two or three nights. (Tried.)

Or, two spoonsful of juice of nettles every morning, and a large cup of decoction of nettles at night, for a week. (Tried.)

Or, three spoonsful of sage juice in a little honey. This presently stops either spitting or vomiting blood. (Tried.)

Or, half a tea-spoonful of Barbadoes tar on a lump of loaf sugar at night. It commonly cures at once.

15. VOMITING BLOOD.

Take two spoonsful of nettle juice.

This also dissolves blood coagulated in the stomach. (Tried.)

Or, take as much saltpeter as will lie upon half-a-crown dissolved in a glass of cold water, two or three times a day.

16. TO DISSOLVE COAGULATED BLOOD.

Bind on the part for some hours a paste of black soap and crumbs of white bread.

Or, grated root of burdock spread on a rag : renew this twice-a-day.

17. BLISTERS.

On the feet, occasioned by walking, are cured by drawing a needleful of worsted through them ; clip it off at both ends and leave it till the skin peels off.

18. BILES.

Apply a little Venice turpentine.
Or, an equal quantity of soap and brown sugar well mixed.
Or, a plaster of honey and wheat flour.
Or, of figs.
Or, a little saffron in a white bread poultice.
'Tis proper to purge also.

19. HARD BREASTS.

Apply turnips roasted till soft, then mashed and mixed with a little oil of roses. Change this twice a day, keeping the breast very warm with flannel.

20. SORE AND SWELLED BREASTS.

*Boil a handful of camomile and as much mallows in milk and water. Foment with it between two flannels, as hot as can be borne, every twelve hours. It also dissolves any knot or swelling in any part.

21. A BRUISE.

* Immediately apply treacle spread on brown paper. (Tried.)
Or, apply a plaster of chopped parsley mixed with butter.
* Or, electrify the part. This is the quickest cure of all.

22. TO PREVENT SWELLING FROM A BRUISE.

Immediately apply a cloth, five or six times doubled, dipped in cold water, and dipped when it grows warm. (Tried)-

23. TO CURE A SWELLING FROM BRUISE.

Foment it half an hour, morning and evening, with cloths dipped in water as hot as you can bear.

24. A BURN OR SCALD.

Immediately plunge the part in cold water: keep it in an hour if not well before. Perhaps four or five hours. (Tried).

* Or, electrify it. If this can be done presently, it totally cures the most desperate burn.

Or, if the part cannot be dipped, apply a cloth four times doubled, dipped in cold water, changing it when it grows warm.

* Or, a bruised onion.

Or, apply oil and strew on it powdered ginger.

25. A DEEP BURN OR SCALD.

Apply black varnish with a feather till it is well.

Or, inner rind of elder well mixed with fresh butter. When this is bound on with a rag, plunge the part into cold water. This will suspend the pain till the medicine heals.

Or, mix lime-water and sweet oil to the thickness of cream, apply it with a feather several times a day. This is the most effectual application I ever met with.

Or, put twenty-five drops of Goulard's extract of lead to half a pint of rain water; dip linen rags in it, and apply them to the part affected. This is particularly serviceable if the burn is near the eyes.

26. A CANCER IN THE BREAST,*

* Of thirteen years standing, was cured by frequently applying red poppy water, plaintain and rose water, mixed with honey and roses. Afterwards, the water used alone perfected the cure.

---

* A cancer is a hard, round, uneven, painful swelling, of a blackish or leaden color, the veins round which seem ready to burst. It comes commonly with a swelling about as big as a pea, which does not at first give much pain, nor change the color of the skin.

Use the cold bath—This has cured many—This cured Mrs. Bates, of LEICESTERSHIRE, of a cancer in her breast, a consumption, a sciatica, and rheumatism, which she had for twenty years. She bathed daily for a month, and drank only water.

A bleeding cancer was cured by drinking twice a day a quarter of a pint of the juice of clivers, or goose grass, and covering the wounds with the bruised leaves.

Another bleeding cancer was cured by the following receipt:

Take half pint of small beer. When it boils, dissolve in an ounce and a half bees' wax: then put in an ounce of hog's lard and boil them together. When it is cold pour the beer from it, and apply it spread on white leather. Renew it every other day. It brings out great blotches, which are to be washed with sal prunello dissolved in warm water.

Monsieur Le Febun advises—"Dissolve four grains of arsnic in a pint of water. *Take a spoonful of this, with a spoonful of milk, and half an ounce of syrup of poppies, every morning.

GENERALLY, where cold bathing is necessary to cure any disease, water drinking is so to prevent a relapse.

If it be not broke, apply a piece of sheet lead beat very thin, and pricked full of pin holes, for days or weeks to the breast. Purges should be added every third or fourth day.

Or, rub the whole breast morning and evening with spirits of hartshorn mixed with oil.

Or, keep it continually moist with honey.

Or take horse spurs† and dry them by the fire, till they

---

†These warts grow on the inside of the horses' fore-legs. *Half a teapoonful at first.

will beat to powder. Sift and infuse two drachms in two quarts of ale; drink half a pint every six hours of new milk warm. It has cured many. (Tried).

Or, a poultice of wild parsnips, flowers, leaves, and stalks, changing it morning and evening; or, scraped carrots.

Or, take brimstone and aqua sulphurata. (See No. 45.) This has cured one far in advance in years. Dr. Cheyne says, a total ass-milk diet, about two quarts a day, without any other food or drink, will cure a confirmed cancer.

27. A CANCER IN ANY OTHER PART.

Apply red onions, bruised.

Or, make a plaster of roche-alum, vinegar, and honey, equal quantities, with wheat flour. Change it every twelve hours. It often cures in three or four days.

Or, stamp the flowers, leaves, and stalks of wild parsnips, and apply them as a plaster, changing it every twelve hours. It usually cures in a few days.

A cancer under the eye was cured by drinking a pint of tar-water daily, washing the same with it, and then applying a plaster of tar and mutton suet melted together. It was well in two months, though of twenty years standing.

28. A CANCER IN THE MOUTH.

Boil a few leaves of succory, plaintain, and rue, with a spoonful of honey, for a quarter of an hour. Gargle with this often in an hour. (I.)

Or, with vinegar and honey, wherein half an ounce of roche-alum is boiled.

Or, mix as much burnt alum and as much black pepper as lies on a sixpence, with an ounce of honey, and frequently touch the part.

Or, blow the ashes of scarlet cloth into the mouth or throat. It seldom fails.

29. CHILBLAINS, (TO PREVENT.)

Wear flannel socks; or socks of chamois leather.

30. CHILBLAINS, (TO CURE.)

Apply salt and onions pounded together.

Or, a poultice of roasted onions hot. Keep it on two or three days if not cured sooner.

Wash them, if broke, with tincture of myrrh in a little water.

31. CHILDREN.

To prevent the rickets, tenderness, and weakness, dip them in cold water every morning, at least till they are nine months old.

No roller should ever be put round their bodies, nor any stays used. Instead of them, when they are put into short petticoats, put a waistcoat under their frocks.

Let them go bare-footed and bare-headed till they are three or four years old at least.

'Tis best to wean a child when seven months old. It should lie in a cradle at least a year.

No child should touch any spiritous or fermented liquors, nor any animal food, before two years old.

Their drink should be water. Tea they should never taste till ten or twelve years old. Milk, milk porridge, and water gruel, are the proper breakfasts for children.

32. CHIN-COUGH, OR HOOPING COUGH.

Use the cold bath daily.

Or, rub the feet thoroughly with hog's lard, before the fire at going to bed, and keep the child warm therein. (Tried.)

Or, rub the back at lying down with old rum. It seldom fails.

Or, give a spoonful of juice of pennyroyal, mixed with brown sugar candy, twice a day.

Or, half a pint of milk warm from the cow, with the quantity of nutmeg or conserve of roses dissolved in it every morning.

Or, dissolve a scruple of salt of tartar in a quarter of a pint of clear water; add to it ten grains of finely powdered cochineal, and sweeten it with loaf sugar.

Give a child within the year the fourth of a spoonful of this four times a day, with a spoonful of barley-water after it. Give a child two years old, half a spoonful; a child above four years old, a spoonful. Boiled apples put into warm milk may be his chief food. This relieves in twenty-four hours, and cures in five or six days.

Or, from three to five grains of gum gamboge. It vomits and purges, and Dr. Cook says, always cures.

Or, take two grains of tartar emetic, and a half a drachm of prepared crab's claws, powdered; let them be mixed very well together.

One grain, one grain and a half, or two grains of this composition may be added to five or six grains of magnesia, and given in a small spoonful of milk and water in the forenoon, between breakfast and dinner, to a child a year old.

At night, if the fever is very high, half the former dose of this powder may be given with from five to ten grains of nitre.

In desperate cases, change of air alone has cured.

33. CHOLERA MORBUS, I. E. FLUX AND VOMITING.

Drink two or three quarts of cold water, if strong; of warm water, if weak.

Or, boil a chicken an hour in two gallons of water, and drink of this till the vomiting ceases.

Or, decoction of rice, or barley, or toasted oaten bread.

If the pain is very severe, stupe the belly with flannels dipped in spirits and water.

The third day after the cure, take ten or fifteen grains of rhubarb.

34. CHOPS IN WOMAN'S NIPPLES.

Apply balsam of sugar.

Or, apply butter and wax, which speedily heals them.

35. CHAPPED HANDS, (TO PREVENT.)

Wash them with flour of mustard.
Or, in bran and water boiled together.

('TO CURE.)

Wash with soft soap mixed with red sand. (Tried.)
Or, wash them in sugar and water. (Tried.)

36 CHAPPED LIPS.

Apply a little sal prunello.

37. A COLD.

Drink a pint of cold water lying down in bed. (Tried.)
Or, to one spoonful of molasses in half a pint of water. (Tried.)
Or, to one spoonful of oatmeal, and one spoonful of honey add a piece of butter the bigness of a nutmeg: pour on, gradually, near a pint of boiling water. Drink this lying down in bed.

38. A COLD IN THE HEAD.

Pare very thin the yellow rind of an orange, roll it up inside out, and thrust a roll into each nostril.

39. THE CHOLIC, (IN THE FIT.)

* Drink a pint of cold water. (Tried.)
Or, a quart of warm water. (Tried.)
Or, of camomile tea.
Or, take from thirty to forty grains of yellow peel of oranges, dried and powdered, in a glass of water.
Or, take from thirty to forty drops of oil of aniseed on a lump of sugar.
Or, apply outwardly a bag of hot oats.
Or, steep the legs in hot water a quarter of an hour.
Or, take as much Daffy's Elixir as will presently purge. This relieves the most violent cholic in an hour or two. See Medicines.

40. THE DRY CHOLIC, (TO PREVENT.)

Drink ginger tea.

41. CHOLIC IN CHILDREN.

Give a scruple of powdered aniseed in their meat. (Tried.)
Or, small doses of magnesia.
Or, put one grain of emetic tartar into four table spoonsful of water : a small tea spoonful will puke a child a week old ; a large tea spoonful is sufficient for one a month old ; and so on in proportion. Repeat the puke every day, or every other day, as the case requires.

This is, perhaps, the best medicine yet discovered for infants. It speedily cures inward fits, gripes, looseness, thrush, and convulsions in children. But if the child is costive, his bowels must be opened first with a little magnesia, or manna, before you give a puke.

42. BILIOUS CHOLIC.*

Drink warm lemonade : I know nothing like it.

Or, give a spoonful of sweet oil every hour. **This has** cured one judged to be at the point of death.

43. AN HABITUAL CHOLIC.

Wear a thin soft flannel on the part.

44. AN HYSTERIC CHOLIC.†

Mrs. Watts, by using the cold bath two and twenty **times** in a month, was entirely cured of an hysteric cholic, fits and convulsive motions, continual sweatings and vomiting, wandering pains in her limbs and head, with total loss of appetite.

In the fit, drink half pint of water with a little wheat flour in it, and a spoonful of vinegar.

Or, of warm lemonade. (Tried.)

---

*This is generally attended with vomiting a greenish or frothy matter, with feverish heat, violent thirst, a bitter taste in the mouth, and little and high colored urine.

† Is attended with a violent pain about the pit of the stomach, **with** great sinking of the spirits and often with greenish vomitings.

Or, take twenty, thirty, or forty drops of balsam of Peru on fine sugar; if need be, take this twice or thrice a day.

Or, in extremity, boil three ounces of burdock seed in water, which give as a clyster.

Or, twenty drops of laudanum in any proper clyster, which gives instant ease.

45. A NERVOUS CHOLIC.*

Use the cold bath daily for three or four weeks.

Or, take quicksilver and aqua sulphurata daily for a month.

46. CHOLIC FROM THE FUMES OF LEAD, OR WHITE LEAD, VERDIGREASE, ETC.†

In the fit, drink fresh melted butter, and then vomit with warm water.

TO PREVENT OR CURE.—Breakfast daily on fat broth, and use oil of sweet almonds frequently and largely.

Smelters of metals, plumbers, etc., may be in a good measure preserved from the poisonous fumes that surround them, by breathing through cloth or flannel mufflers twice or thrice doubled, dipped in a solution of sea salt, or salt of tartar, and then dried. These mufflers might also be of great use in similar cases.

47. WINDY CHOLIC.

Parched peas eaten freely, have had the most happy effect when all other means have failed.

---

* A cholic with purging, some term the watery gripes.

† This some term the dry belly-ache. It often continues several days, with little urine and obstinate costiveness.

48. TO PREVENT THE ILL EFFECTS OF COLD

The moment a person gets into a house with his hands or feet chilled, let him put them into a vessel of water, as cold as can be got, and hold them there till they begin to glow. This they will do in a minute or two. This method likewise effectually prevents chilblains.

49. A CONSUMPTION.

Cold bathing has cured many deep consumptions. (Tried.)
One in a deep consumption was advised to drink nothing but water, and eat nothing but water gruel, without salt or sugar. In three months time he was perfectly well.

Take no food but new buttermilk, churned in a bottle, and white bread. I have known this successful.

Or, use as a common drink, spring water and new milk, each a quart, and sugar candy two ounces.

Or, boil two handsful of sorrel in a pint of whey, strain it, and drink a glass thrice a day. (Tried).

Or, turn a pint of skimmed milk with half a pint of small beer. Boil in this whey about twenty-five ivy leaves, and two or three sprigs of hyssop. Drink half over night, the rest in the morning. Do this, if needful, for two months daily. This has cured in a desperate case. (Tried.)

Or, take a cow heel from the tripe house ready dressed, two quarts of new milk, two ounces of hartshorn shavings two ounces of isinglass, a quarter of a pound of sugar candy, and a race of ginger. Put all these in a pot; and set them in an oven after the bread is drawn. Let it continue there till the oven is near cold; and let the patient live on this. I have known this cure a deep consumption more than once.

Or every morning, cut a little turf of fresh earth, and laying down, breathe in the hole for a quarter of an hour. I have known a deep consumption cured thus.

"Mr. Masters, of Evesham, was so far gone in a consumption that he could not stand alone. I advised him to lose six ounces of blood every day for a fortnight, if he lived so long; and then every third day; then every fifth day for the same time. In three months he was well.—*Doctor Dover*. Tried.

Or, throw frankincense on burning coals, and receive the smoke daily through a proper tube into the lungs. (Tried.)

Or, take in for a quarter of an hour, morning and evening, the steam of white resin and bees' wax boiling on a hot fire shovel. This has cured one who was in the third stage of consumption.

Or, the steam of sweet spirit of vitriol dropped into warm water.

Or, take, morning and evening, a tea spoonful of white resin powdered and mixed with honey. This cured one in less than a month, who was near death.

Or, drink thrice a day two spoonsful of juice of water-cresses. This has cured a deep consumption.

In the last stage suck a healthy woman daily. This cured my father.

For diet, use milk and apples, or water gruel made with fine flour. Drink cider whey, barley water, sharpened with lemon juice or apple water.

So long as the tickling cough continues, chew well, and swallow a mouthful or two, of a biscuit or crust of bread twice a day. If you cannot swallow it spit it out. This will always shorten the fit, and would often prevent consumption.

50 CONVULSIONS.

Use the cold bath.

Or, take a tea spoonful of valerian root powdered, in a cup of water every evening.

Or, half a drachm of misselto powdered, every six hours drinking after it a draught of strong infusion thereof.

51. CONVULSIONS IN CHILDREN.

Scrape piony roots fresh digged. Apply what you have scraped off to the soles of the feet. It helps immediately. (Tried).

52. CONVULSIONS IN THE BOWELS OF CHILDREN.

Give a child a quarter old, a spoonful of the juice of *pellitory of the wall, two or three times a day; it goes through at once, but purges no more. Use the syrup if the juice cannot be had.

---

* An European plant. See dispensatory.

53 CORNS, (TO PREVENT).

Frequently wash the feet in cold water.

54 CORNS, (TO CURE.)

Apply fresh, every morning, the yeast of small beer spread on a rag.

Or, after paring them close, apply bruised ivy leaves daily, and in fifteen days they will drop out. (Tried.)

Or, apply chalk powdered and mixed with water. This also cures warts.

Some corns are cured by a pitch plaster.

All are greatly eased by steeping the feet in hot water wherein oat meal is boiled. This also helps dry and hot feet.

55. COSTIVENESS.

Rise early every morning.

Or boil in a pint and a half of broth, half a handful of mallow leaves chopped, strain this and drink it before you eat anything else. Do this frequently, if needful.

Or, breakfast twice a week or oftener, on water gruel with currants. (Tried.)

Or, take the bigness of a large nutmeg of cream of tartar mixed with honey, as often as you need.

Or, take daily, two hours before dinner a small tea cupful of stewed prunes.

Or, use for common drink, water or treacle beer, impregnated with fixed air.

Or, live upon bread made of wheat flour with all the bran in it.

Or, boil an ounce and a half tamarinds in three pints of water to a quart. In this strained, when cold, infuse all night two drachms of senna, and one drachm of red rose leaves, drink a cupful every morning.

56. A COUGH.

Every cough is a dry cough at first. As long as it continues so, it may be cured by chewing immediately after you cough, the quantity of a pepper corn of Peruvian bark. Swallow your spittle as long as it is bitter, and spit out the wood. If you cough again, do this again. It very seldom fails to cure any dry cough. I earnestly desire every one, who has any regard for his health, to try this within twenty-four hours after he perceives a cough.

Or, drink a pint of cold water lying down in bed. (Tried.)

Or, make a hole through a lemon and fill it with honey. Roast it and catch the juice. Take a tea spoonful of this frequently. (Tried.)

57. AN ASTHMATIC COUGH.

Take Spanish liquorice two ounces, salt of tartar half an ounce; boil the liquorice in three pints of water to a quart: add the salt to it when it is blood warm. Drink two spoonsful of this every two hours. It seldom fails. (Tried.) I have known this cure an inveterate moist asthma.

58. A CONSUMPTIVE COUGH.

To stop it for a time, at lying down keep a little stick liquorice shaved like horse radish, between the cheek and gums. I believe this never fails.

59. A CONVULSIVE COUGH.

Eat preserved walnuts.

60. AN INVETERATE COUGH.

Wash the head in cold water every morning.
Or, use the cold bath. It seldom fails.
Or, peel and slice a large turnip, spread coarse sugar between the slices, and let it stand in a dish till all the juice drains down. Take a spoonful of this when you cough.

Or, take a spoonful of syrup of horehound morning and evening. (Tried.)

Or, take from ten to twenty drops of elixir of vitriol, in a glass of water, twice or thrice a day. This is useful when the cough is attended with costiveness, or a relaxation of the stomach and lungs.

61. A PLEURITIC COUGH.

Powder an ounce of spermaceti fine. Work it in a mortar with the yolk of a new laid egg. Mix them in a pint of of white wine. and take a small glass every three hours.

62. A TICKLING COUGH.

Drink water whitened with oatmeal four times a day.

Or, keep a piece of barley sugar or sugar candy constantly in the mouth.

63. VIOLENT COUGHING FROM A SHARP AND THIN RHEUM.*

Work into old coserve of roses as much as you can of pure frankincense powdered as fine as possible. Take a bolus of this twice or thrice a day.

Or, take half a grain of inspissated milky juice of sow thistle, once or twice a day. It has the anodyne and antispasmodic properties of opium without its narcotic effects. Or it may be made into laudnum in the same manner as opium is, and five or six drops taken on a lump of sugar thrice a day.

The milky juice of all the sowthistles, dandelions, and lettuces, have nearly the same virtues.

Or, use milk diet as much as possible.

64. THE CRAMP, (TO PREVENT.)

Tie your garter smooth and tight under your knee at going to bed : it seldom fails.

Or, take a half a pint of tar water morning and evening.

Or, be electrified through the part which uses to be affect-

---

* A thin fluid secreted by the glands.

ed. This generally prevents it for a month, sometimes for a twelvemonth.

Or, to one ounce and a half of spirits of turpentine, add flour of brimstone and sulphur vivum, of each half an ounce, smell to it at night three or four times.

Or, lay a roll of brimstone under your pillow.

65. THE CRAMP, (TO CURE.)

Strongly put out your heel.
Or, chafe the part with Hungary water.
Or, hold a roll of brimstone in your hand. I have frequently done this with success.

66. A CUT.

Keep it closed with your thumb for a quarter of an hour, Then double a rag five or six times, dip it in cold water, and bind it on. (Tried.)

Or, bind on toasted cheese. This will cure a deep cut.

Or, pounded grass. Shake it off after twelve hours, and, if need be, apply fresh.

67. DEAFNESS.

Be electrified through the ear. (Tried.)
Or, use the cold bath.
Or, put a little salt into the ear.

Or, drop into it a tea spoonful of salt water.

Or, three or four drops of onion juice at lying down, and stoped in with a little wool.

68. DEAFNESS FROM WAX.

Syringe the ear with warm water. (Tried.)

69. DEAFNESS WITH A DRY. EAR.

Mix brandy and sweet oil; dip black wool in this and put it into the ear. When it grows dry, wash it well in brandy, dip it and put it in again.

70. DEAFNESS WITH A HEADACHE AND BUZZING IN THE HEAD.

Peel a clove of garlic, dip it in honey, and put it into your ear at night with a little black wool. Lie with that ear uppermost. Do this, if need be, eight or ten nights (Tried.)

71. A SETTLED DEAFNESS.

Take a red onion, pick out the core, and fill up the place with oil of roasted almonds. Let it stand a night then bruise and strain it. Drop three or four drops into the ear morning and evening, and stop it with black wool.

72. DELIVERY.

After delivery in child birth, the mother's milk is the only proper purge for the child. Let it begin to suck ten or twelve hours after the birth.

73. A DIABETES.‡

Drink wine, boiled in ginger, as much and as often as your strength will bear. Let your drink be milk and water. All milk meats are good.

Or, drink three or four times a day a quarter of a pint of alum posset, putting three drachms of alum to four pints of milk. It seldom fails to cure in eight or ten days.—*Dr. Mead.*

* Or, infuse half an ounce of cantharides in a pint of elixir of vitriol. Give from fifteen to thirty drops in Bristol water twice or thrice a day.

74. THE DROPSY.†

Use the cold bath daily, after purging.

Or, rub the swelled parts with salad oil by a warm hand,

---

‡ A diabetes is a frequent and large discharge of sweetish urine, attended with a constant thirst aud a wasting of the whole body.

*Use with caution.

† A dropsy is a preturnatural collection of water in the head, breast, belly, or all over the body. It is attended with a continual thirst. The part swelled pits if you press it with your fingers. The urine is pale and little.

at least one hour a day. This has done wonders in some cases.

Or, cover the whole belly with a large new sponge dipped in strong lime water, and then squeezed out. This bound on often cures, even without any sensible evacuation of water.

Or, apply green dock leaves to the joints and soles of the feet, changing them once a day.

Or, mix half an ounce of amber with a quart of wine vinegar. Heat a brick, only not red hot, and put it into a tub. Pour them upon it and hold the part swelled over the smoke, covering the tub close to keep in the smoke. The water will come out incredibly, and the patient be cured. (Tried).

Or, eat a crust of bread every morning, fasting. (Tried.)

Or, take as much as lies upon a sixpence of powdered laurel leaves, every second or third day. It works both days. (Tried.)

Or, mix a pound of the coarsest sugar with a pint of juice of pellitory of the wall, bruised in a marble mortar. Boil it as long as any scum rises. When cool, bottle and cork it. If very bad, take three spoonsful at night and two in the morning. It seldom fails. (Tried.)

Or, make tea of roots of dwarf elder. It works by urine. every twelve or fourteen minutes (that is, after each discharge) drink a tea cupful. I have known a dropsy cure by this in twelve hours time.

One was cured by taking a drachm of nitre, every morning, in a little ale.

Tar water drank twice a day has cured many, so has an

infusion of juniper berries, roasted, and made into a liquor like coffee.

Or, three spoonsful of the juice of leeks or elder leaves. (Tried.) This cured the windy dropsy.

Or, half a pint of decoction of butcher's broom, (intermixing purges twice or thrice a week.) The proper purge is ten grains of jalap with six of powdered ginger. It may be increased or lessened according to the strength of the patient.

Or, of the decoction of the tops of oak boughs. This cured an inveterate dropsy in fifteen days.

Or, take senna, cream of tartar, and jalap, half an ounce each. Mix them and take a drachm every morning in broth. It usually cures in twenty days. This is nearly the same with Dr. Ward's powder : I suppose he took it from hence. He says it seldom fails, either in the watery or windy dropsy.

Or, be electrified. This cures dropsies supposed to be incurable.

How AMAZINGLY little is yet known, even of the human body ! Have not dropsical persons been continually advised to abstain from drink as much as possible ? but how can we reconcile this with the following undeniable facts, published in the Medical Transactions.

Jane Roberts, aged twenty, was at length constrained to take her bed by a confirmed ascites and anasarca. In this desperate case she drank as much as she would, first of small beer, and when that failed, of thin milk. After awhile her skin cracked in many places, and she continued drinking and leaking till she was quite well.

A middle-aged man in the west of England, drank every day five or six quarts of cider; and, without any other medicine, was totally cured in a few weeks of a dropsy long supposed to be incurable.

A farmer, aged seventy, in a confirmed ascites was given over for dead. Being desperate, he drank three quarts of cold water every four and twenty hours. His whole food in the meantime was sea biscuit, sometimes with a little butter For sixteen days he seemed worse. Then he discharged, for near a week, a vast quantity of water, and was soon free from his disease, which never returned.

75. DROWNED.

Rub the trunk of the body all over with salt. It frequently recovers them that seem dead.

76. THE EAR ACHE.

Rub the ear hard for a quarter of an hour. (Tried.)
Or, be electrified.
Or, put in a roasted fig or onion, as hot as may be. (Tried.)
Or, blow the smoke of tobacco strongly into it.
But if the ear ache is caused by an inflammation of the uvula, it is cured in two or three hours by receiving into the mouth the steam of bruised hemp seed boiled in water.

77. EAR ACHE FROM COLD.

Boil rue, or rosemary, or garlic, and let the steam go into the ear through a funnel.

78. EAR ACHE FROM HEAT.

Apply cloths four times doubled and dipped in cold water changing them when warm for an half an hour.

79. EAR ACHE FROM WORMS

Drop in warm milk, which brings them out.
Or, juice of wormwood, which kills them.

80. NOISE IN THE EARS.

Drop in juice of onions.

81. HARD WAX IN THE EARS;

Is best dissolved by syringing the ears with warm water

82. EYES BLEARED.

Drop into them juice of crab apples.

83. BLOOD-SHOT EYE.

Apply linen rags dipped in cold water for two or three hours.
Or, blow in white sugar candy, finely powdered.
Or, apply boiled hyssop as a poultice. This has a wonderful efficacy.

84. A BRUISE IN THE EYE.

Apply a plaster of the concerve of roses.

85. CLOUDS FLYING BEFORE THE EYE.

Take a drachm of powdered betony every morning.
Or, be electrified.

86. BLINDNESS.

It is often cured by cold bathing.
Or, by electrifying. (Tried.) This has cured a suffusion of sixteen, and a gutta serena of twenty-four years standing.

87. DULL SIGHT.

Drop in two or three drops of juice of rotten apples often.

88. FILMS.

Dry zibethum occidentale, i. e. stercus humanum, slowly; powder it fine, and blow it into the eye twice or thrice a day.

Or, mix juice of ground-ivy with a little honey and two or three grains of bay salt. Drop it in morning and evening.

Or, touch them cautiously every day with the lunar caustic.

89. HOT OR SHARP HUMORS.

Apply a few drops of double refined sugar melted in brandy. (Tried).

Or, boil a handful of bramble leaves with a little alum in a quart of spring water, to a pint. Drop this frequently into the eye. This likewise speedily cures cancers or any sores.

Or, lay a thin slice of raw beef on the nape of the neck (Tried.)

90. EYES OR EYE LIDS INFLAMED.

Apply as a poultice, boiled, roasted, or rotten apples warm.

Or, wormwood tops with the yolk of an egg. This will hardly fail.

Or, beat up the white of an egg with two spoonsful of white rose water into a white froth. Apply this on a fine rag, changing it so that it may not grow dry till the eye or eye-lid is well. (Tried.)

Or, dissolve an ounce of fine gum arabic in two or three spoonsful of spring water; put a drop into the inner corner of the eye, from the point of a hair pencil, four or five times a day. At the same time take as much saltpeter as will lie upon a sixpence, dissolved in a glass of water, three or four times a day ; abstaining from all liquors as much as possible till cured. White bread poultices applied to the eyes in an inflamed state, frequently occasions total blindness.

After the inflammation is subsided, if weakness still remains, dip a finger in the white copperas eye-water and rub round the eye three or four times a day.

N. B. All acrid eye-water and powders, put into the eyes when they are much inflamed, horribly increase both the pain and inflammation.

91. LACHRYMAL FISTULA,*

Apply a poultice of fine leaves of rue.

Or, wash the eye morning and evening with a decoction of quince leaves.

92. PEARL IN THE EYE.

Apply a drop of juice of calendine with a feather thrice a day.

Or, of three-leaved grass. It commonly cures in seven days.

Or, dissolve a little sal ammoniac in rose water. Keep this three days in a copper vessel. Drop it twice a day into the eye.

Or, reduce separately, to the finest powder possible, an

---

*This disorder in the inner corner of the eye, causes the tears to flow involuntarily. When it is confirmed, only a surgeon can cure it.

equal weight of loaf sugar, cream of tartar, and bole armenia; mix them together, and put a little into the eye, without blowing it in, three or four times a day.

93. SORE EYES.

Drink eye-bright tea, and wash the eyes with it.

94. WHITE SPECKS IN THE EYES.

Going to bed, put a little ear-wax on the specks. This has cured many.

95. AN EXCELLENT EYE-WATER.

Put half an ounce of lapis calaminaris powdered, into half a pint of French white wine and as much white rose-water; drop a drop or two into the corner of the eye. I have known it cure total blindness.

96. ANOTHER.

Boil very lightly one spoonful of white copperas, scraped, and three spoonsful of white salt, in three pints of spring water. When cold, bottle it in large phials without straining. Take up a phial softly, and put in a drop or two in the eye morning and evening.

It answers the intention of almost all the preceding medicines, and takes away redness or any soreness whatever; it cures pearls, rheums, and often blindness itself. But if it makes the eye smart, add more water to it.

97. ANOTHER.

Stamp and strain ground-ivy, calendine, and daisies, an equal quantity: add a little rose water and loaf sugar. Drop a drop or two at a time into the eye, and it takes away all manner of inflammation, smarting, itching, spots, webs, or any other disorder whatsoever, yea, though the sight were almost gone.

98. AN EYE-WATER,

Which was used by Sir Stephen Fox, when he was sixty years of age, and could hardly see with the help of spectacles; but hereby in some time he recovered his sight, and could read the smallest print, without spectacles, till above eighty.

Take six ounces of rectified spirits of wine, dissolve in it one drachm of camphire, then add two small handsful of dried elder flowers. In twenty-four hours after it is infused, it is ready for use. Take out a little on a tea-spoon, dip your finger in it and bathe your forehead over your eyes and each temple with it several times, morning and night, and twice more in the day constantly. Meantime dip a soft rag in dead small beer, new milk warm, and dab each eye a dozen times, gently, morning and evening.

If it is a watery humour, you may with your finger wet the

eye lids two or three times a piece ; but be sure to shut you eyes, or it makes them smart and burn excessiveiy. If you have the tooth-ache, or swelled face, rub it well on the part and it will take away the pain. It will cure any bruise also if used immediately. (Tried.)

99. WEAK EYES.

Wash the head daily with cold water. (Tried.)

100 FAINTING ON LETTING BLOOD,

Is prevented by taking before it some good broth.

101. THE FALLING SICKNESS.*

Be electrified. (Tried.)

Or, use the cold bath for a month, daily.

Or, take a tea-spoonful of piony root, dried and grated fine, morning and evening for three months.

Or, half a spoonful of valerian root powdered, three times a day, in a glass of water, for three months.

Or, half a pint of tar water, morning and evening, for three months.

Or, a glass of juice of pellitory of the wall, every morning.

---

*In the falling sickness, the patient falls to the ground either quite stiff or convulsed all over, utterly senseless, gnashing his teeth, and foaming at the mouth,

Or, take five or six drops of laudanum fasting, for six or seven mornings. This has cured many.

Or, use an entire milk diet for three months. It seldom fails.

In the fit, blow up the nose a little powdered ginger. Or, leaves of assarabacca powdered.

THIS is the famous Major's snuff.

Or, blow down the throat the smoke of tobacco.

* One who is subject to the falling sickness, may prevent the fit if he feels it coming, by this simple experiment: let him always carry with him a piece of metal as broad as he is able to hold between his teeth, when his jaws are stretched to the utmost. When he feels the fit approaching, let him immediately put this between his teeth so as to keep his jaws at the utmost stretch. In about a minute this will bring him quite to himself, and prevent the fit for that time.

If one put this metal between the teeth of one that is in the fit, and force them open till his jaws are at the utmost stretch, the fit will immediately go off, and the patient very soon recover.

102. THE FALLING OF THE FUNDAMENT.

Apply a cloth covered thick with brick-dust.

Or, boil a handful of red rose leaves in a quarter of a pint of red wine, dip a cloth in it, and apply it as hot as can be borne. Do this till all is used.

103. A FALLING DOWN OF THE WOMB

May be cured in the manner last mentioned.

Or, wear a pessory of cork, and take once or twice a day a tea-cupful of the decoction of the bark, with ten drops of the elixir of vitriol.

104. EXTREME FAT.

Use a total vegetable diet. I knew one who was entirely cured of this by living a year thus: she breakfasted and supped on milk and water (with bread) and dined on turnips, carrots, or other roots, drinking water.

105. A FEVER.

In the beginning of any fever, if the stomach is uneasy, vomit; if the bowels purge; if the pulse be hard, full, or strong, bleed.*

Drink a pint or two of cold water lying down in bed; I never knew it to do hurt.

Or, a large glass of tar-water, warm every hour.

Or, thin water gruel sweetened with honey, with one or two drachms of nitre in each quart.

THE best of all juleps in a fever is thus: Toast a large thin slice of bread, without burning; put it hot into a pint of cold water, then set it on the fire till it is pretty hot. In a dry heat it may be given cold; in a moist heat, warm; the more largely the better. (Tried.)

Or, for a change, use pippin or wood-sorrel tea; or pippin whey, or wood-sorrel whey.

---

*We seldom advise bleeding. For substitute see second part of this book.

(To prevent catching any infectious fever, do not breathe near the face of the sick person, neither swallow your spittle whilst in the room. Infection seizes the stomach first.)

Or, stamp a handful of leaves of woodbine, put soft water to it, and use it cold as a clyster. It often cures in an hour.

Or, smear the wrists five or six inches long with warm molasses, and cover it with brown paper.

Or, apply molasses plasters to the head and the soles of the feet, changing them every twelve hours.

Or, use Dr. Boerhaave's fever powder, viz: Eight ounces of nitre, a quarter of an ounce of camphire, half a quarter of an ounce of saffron, and eight grains of cochineal. These are to be powdered, mixed together, and kept dry in a bottle. Ten grains taken at going to bed abate feverish heat, and procure rest. Ten grains are to be taken every three or four hours for a continued fever.

106. A HIGH FEVER,

Attended with a delirium and a vigilia, has been cured by plunging into cold water; which is a safe and sure remedy* in the beginning of any fever.

Such a delirium was often cured by applying to the top of the head a molasses plaster. (Tried).

107. AN INTERMITTING FEVER.

Drink warm lemonade in the beginning of every fit; it cures in a few days. (Tried.)

---

*Only *robust* persons should use this remedy.

Or, take a tea-spoonful of oil of sulphur in a cup of balm tea, once or twice a day.

108. A FEVER WITH PAINS IN THE LIMBS.

Take twenty drops of spirit of hartshorn in a cup of water twice or thrice in twenty-four hours.
Or drink largely of cinque-foil tea.

109. A RASH FEVER.

Drink every hour a spoonful of juice of gronnd-ivy. It often cures in twenty-four hours. Use this decoction when you have not the juice.

110. A SLOW FEVER.

Use the cold bath for two or three weeks, daily.

111. A WORM FEVER.

Boil a handful of rue and wormwood in water; foment the belly with the decoction, and apply the boiled herbs as a poultice; repeat the application night and morning. This frequently brings away worms from children who will take no internal medicine, and is likewise serviceable if the fever be of the putrid kind.

112 A FISTULA.

Wash muscle shells clean, burn them to powder, sift them

fine, mix them with hog's lard, spread it on clean wash leather, and apply it. This cured one who was thought to be at the point of death.

N. B. This also cures the piles.

Or, grind an ounce of sublimate mercury in a glass pestile, as fine as possible. Put it into a glass bottle, and pour on it two quarts of pure spring water.
Cork it close, and for six days shake it well every hour. Then let it settle for twenty-four hours. Pour it off clear, filter in a glass funnel and keep it for use, close stopped.

[Should be kept out of the reach of children and labeled poison. For external use as below.—ED.]

Or, have a vessel so contrived, that you may sit with the part in cold water a quarter of an hour, every morning. I have known a gentleman of seventy cured thereby.

Or, put a large stone of unslacked lime into four quarts of water, let it stand one night; take four ounces of roche-alum, and four ounces of white coperas, calcine them to dryness, then powder them as fine as possible: take three pints of the above water and put the powder into it and boil it for an half hour, then let it cool, and bottle it for use. Let the fistula be syringed with this often a little warm; and apply it twice a day. Cover with a plaster of diaculum.

This water will destroy the callosity of the edges of the fistula, which otherwise would prevent its healing, and managed as above, will heal it up at the same time.

113. TO DESTROY FLEAS AND BUGS.

Cover the floor of the room with leaves of black alder

gathered while the dew hangs upon them : adhering to these they are killed thereby.

Or, powder stavesacre and sprinkle it on the body, or on the bed.

114. PHLEGM.

To prevent, or cure, take a spoonful of warm water first thing in the morning.

115. FLOODING, (IN LYING IN.)

Cover the body with cloths dipped in vinegar and water changing them as they grow warm. Drink cooling acid liquors.

This is a complaint which is not to be thought little of. Sometimes a violent flooding comes on before delivery ; and the only way to save both the mother and the child is to deliver the woman immediately, which, being done, the flooding will generally cease. Sometimes a slight flooding comes on some weeks before labor ; and here if the patient be kept cool, her diet light, and small doses of nitre often repeated, (an ounce divided into thirty parts, and one given every four hours) she will frequently go her full time and do well; but if it should become excessive, delivery should be effected as soon as may be.

If a flooding should come on after delivery, the patient should be laid with her head low, kept cool, and be in all respects treated as for an excessive flux of the menses. Linen cloths which have been wrung out of vinegar and water, should be applied to the belly, the loins, and the thighs :

These must be changed as they grow dry, may be discontinued as soon as the flooding abates. Sometimes the following mixture will be very useful, viz: Pennyroyal-water, simple cinnamon-water, and syrup of poppies, of each two ounces; acid elixir of vitriol, one drachm. Mix and take two table-spoonful every hour. But large doses of nitre given often (a scruple every hour) is generally the most efficacious. But when all other things seem to have no effect cold water dashed upon the patient's belly will stop the flooding immediately.

116. A FLUX.

Receive the smoke of turpentine cast on burning coals This cures also the bloody flux, and the falling of the fundament.

Or put a large brown toast into three quarts of water, with a drachm of cochineal, powdered, and a drachm of salt of wormwood. Drink it all in as short a time as you conveniently can.

This rarely fails to cure all fluxes, cholera morbus, yea and inflammation of the bowels. Tried.

Or, take a spoonsful of plantaine seed, bruised, morning and evening, till it stops.

Or, ten grains of ipecacuanha, three mornings successively It is likewise excellent as a sodorific.

Or, boil four ounces of rasped logwood, or fresh logwood chips, in three quarts of water to two; strain it, and drink a quarter of a pint sweetened with loaf sugar, warm, twice a day. It both binds and heals. Or, take a small tea-cupful of it every hour.

Or, boil the fat of a breast of mutton in a quart of spring water for an hour. This will cure the most inveterate flux. (Tried.

117. A BLOODY FLUX.

Apply suppository of linen dipped in aqua vitæ.

Or, drink cold water as largely as possible, taking nothing else till the flux stops.

Or, take a large apple, and at the top pick out all the core, and fill up the place with a piece of honey-comb, the honey being strained out; roast the apple in embers and eat it, and this will stop the flux immediately.

Or, grated rhubarb, as much as lies on a shilling, with half as much of grated nutmeg, in a glass of white wine, at lying down every other night. (Tried.)

Or, take four drops of laudanum, and apply to the belly a poultice of wormwood and red roses, boiled in milk.

In a dysentery, the worst of all fluxes, feed on rice, saloup, sago, and sometimes on beef tea; but no flesh.

To stop it, take a spoonful of suet melted over a slow fire. Do not let blood.

A PERSON was cured in one day by feeding on rice milk, and sitting a quarter of an hour in a shallow tub, having in it warm water, three inches deep.

118. GANGRENE, (TO PREVENT OR STOP A BEGINNING.)

Foment continually with vinegar, in which dross of iron (either sparks or clinkers) has been boiled.

119. THE GOUT IN THE STOMACH.

Dissolve two drachms of Venice treacle in a glass of mountain wine. After drinking it go to bed. You will be easier in two hours, and well in sixteen.—*Dr. Dover.*

Or, boil a *pugil of tansey in a quarter of a pint of mountain wine. Drink it in bed. I believe this never fails

To prevent its return, dissolve half an ounce of gum-guiacum in two ounces of sal volatile. Take a tea-spoonful of this every morning in a glaas of spring water.

This helps any sharp pain in the stomach—*Dr. Boerhaave.*

N. B. I knew a gentleman who was cured many times by a large draught of cold water.

120. THE GOUT IN THE FOOT OR HAND.

Apply a raw lean beef steak. Change it once in twelve hours till cured. (Tried.)

121. THE GOUT IN ANY LIMB.†

Rub the part with warm treacle, and then bind on a flannel smeared therewith. Repeat this it need be, once in twelve hours.

---

*A pugil is as much as you can take up between your thumb and two fingers.

†Regard not them who say the gout ought not to be cured; they mean it cannot. I know it cannot by their regular prescriptions; but I have known it cured in many cases without any ill effects following.

THIS has cured an inveterate gout in thirty-six hours.

Or, drink a pint of strong infusion of elder buds dry or green, morning and evening. This has cured inveterate gouts.

Or, at six in the evening, undress and wrap yourself up in blankets; then put your legs up to the knees in water, as hot as you can bear it. As it cools, let hot water be poured in so as to keep you in a strong sweat till ten. Then go to bed well warmed, and sweat till morning. I have known this cure an inveterate gout in a person above sixty, who lived eleven years after. The very matter of the gout is frequently destroyed by a steady use of Mynsicht's elixir of vitriol.*

122. THE GRAVEL.

Eat largely of spinach.

Or, drink largely of warm water sweetened with honey.

Or, of pellitory of the wall tea so sweetened.

Or, infuse an ounce of wild parsley-seeds in a pint of white wine for twelve days. Drink a glass of it, fasting, three months. To prevent its return, breakfast for three months on agrimony tea. It entirely cured me twenty years ago, nor have I had the least symptom of it since.

123. GREEN SICKNESS.

Take a cup of decoction of lignum guiacum (commonly called lignum vitæ) morning and evening.

---

* Elixir of vitriol of the dispensatory practicaly the same.

Or, grind together into a fine powder, three ounces of the finest steel finings, and two ounces of red sugar candy. Take from a scruple to half a drachm every morning. (I.)

124. TO KILL ANIMALCULA THAT CAUSE THE GUMS TO WASTE AWAY FROM THE TEETH.

Gargle thrice a day with salt and water.

125. TO MAKE HAIR GROW.

Wash it every night with a strong decoction of rosemary. Dry it with flannel. (Tried.)

126. THE HEAD-ACHE.

Rub the head for a quarter of an hour. (Tried.)
Or, be electrified. (Tried.)
Or, apply to each temple the thin yellow rind of a lemon newly pared off.
Or, pour upon the palm of the hand a little brandy and some zest* of lemon, and hold it to the forehead; or a little æther.
Or, if you have catched cold, boil a handful of rosemary in a quart of water. Put this in a mug, and hold your head, covered with a napkin, over the steam as hot as you can bear it. Repeat this till the pain ceases. (Tried.)

---

* Zest is the juice of the peel squeezed out.

Or, snuff up the nose camphorated spirits of lavender.
Or, a little of horse-radish.

127. A CHRONIC HEAD-ACHE.

Keep your feet in warm water a quarter of an hour before you go to bed, for two or three weeks. (Tried.)

Or, wear tender hemlock leaves under the feet, changing them daily.

Or, order a tea-cupful of cardnus tea without sugar, fasting, for six or seven mornings. (Tried.)

128. HEAD-ACHE FROM HEAT

Apply to the forehead cloths dipped in cold water.

129. A NERVOUS HEAD-ACHE.

Dry and powder an ounce of marjoram and half an ounce of assarabacca; mix them as snuff, keeping the ears and throat warm. This is of great use even in a cancer; but it will suffice to take a small pinch every other night, lying down in bed.

130. A VIOLENT HEAD-ACHE.

Take of white-wine vinegar and water, each three spoonsful, with half a spoonful of Hungary water. Apply this twice a day to the eyelids and temples.

131. HEMICRANIA.*

Use cold bathing.
Or, apply to that part of the head shaven, a plaster that will stick, with a hole cut in the middle of it as big as a half penny; place over that hole the leaves of ranunculus† bruised and very moist. It is a gentle blister.

132. STOPPAGE IN THE HEAD.

Snuff up juice of primrose, keeping the head warm.

133. THE HEART BURNING.‡

Drink a pint of cold water. (Tried.)
Or, drink slowly decoction of camomile flowers.
Or, eat four or five oysters.
Or, chew five or six pepper corns a little, then swallow them.
Or, chew fennel or parsley, and swallow your spittle.

---

* This is a head-ache which affects but one side of the head.
† Crow-foot.
‡ A sharp gnawing pain in the orfice of the stomach.

Sometimes a vomit is needful.
Or, a piece of Spanish liqorice.

134. THE HICCUP, (TO PREVENT.)

Infuse a scruple of musk in a quart of mountain wine, and take a small glass every morning.

(TO CURE.)

Swallow a mouthful of water, stopping the mouth and ears. (Tried.)

Or, take anything that will make you sneeze.

Or, three drops of oil of cinnamon on a lump of sugar. (Tried).

Or, two or three preserved damsons.

Or, ten drops of chemical oil of amber dropped on sugar, and then mixed with a little water.

135. HOARSENESS.

Rub the soles of the feet, before the fire, with garlic and lard well beaten together, over night. The hoarseness will be gone next morning. (Tried.)

Or, take a pint of cold water lying down.

Or swallow slowly the juice of radishes.

Or, half a pint of mustard whey lying down.

Or, a tea-spoonful of conserve of roses every night. (Tried

Or, dry nettle roots in an oven; then powder them finely,

and mix with an equal quantity of molasses. Take a teaspoonful twice a day.

Or, boil a large handful of wheat bran in a quart of water, strain, and sweeten it with honey. Sup of it frequently.

136. HYPOCONDRIAC AND HYSTERIC DISORDERS.

Use cold bathing.

Or, take a physic nearly every morning, and ten drops of elixir of vitriol in the afternoon, in a glass of water.

131. THE JAUNDICE.

Wear leaves of calendine upon and under the feet.

Or, take a small pill of Castile soap every morning for eight or ten days. (Tried.)

Or, beat the white of an egg thin; take it morning and evening in a glass of water. (I.)

Or, half a pint of strong decoction of nettles, or of burdock leaves morning and evening.

Or, boil three ounces of burdock root, in two quarts of water to three pints. Drink a tea-cupful of this evey morning.

138. JAUNDICE IN CHILDREN.

Take half an ounce of fine rhubarb, powdered, mix with it

thoroughly, by long beating, two handsful of good well cleansed currants. Of this give a tea-spoonful every morning.

139. THE ILIAC PASSION.*

Apply warm flannels soaked in spirits of wine.

Or, hold a live puppy constantly on the belly.—*Dr. Sydenham.*

Inflammations in general are more certainly abated by smart purging than by bleeding.

140. AN IMPOSTHUME †

Put the white of two leeks in a wet cloth, and so roast them in ashes, but not too much. Stamp them in a mortar with a little hog's grease. Spread it thick plaster-wise, and apply it, changing it every hour till all the matter is come out, which will be in three times. (I.)

141 THE ITCH.‡

Wash the parts affected with strong rum. (Tried).

---

*In this violent kind of cholic the execrements are supposed to be thrown up by the mouth in vomiting.

† A forming abcess.

‡ This distemper is nothing bur a kind of very small lice which burrow under the skin ; therefore inward medicines are absolutely needless. Is it possible any physician should be ignorant of this ?

Or, anoint them with black soap, but wash it off soon.

Or, steep a shirt half an hour in a quart of water mixed with half an ounce of powdered brimstone. Dry it slowly and wear it five or six days. Sometimes it needs repeating. (Tried.)

Or, mix powder of white hellebore with cream for three days. Anoint the joints for three mornings and evenings. It seldom fails.

Or, beat together the juice of two or three lemons, with the same quantity of oil of roses. Anoint the parts affected. It cures in two or three times using.

142. THE KING'S EVIL.*

Take as much cream of tartar as lies on a sixpence, every morning and evening.

Or, drink for six weeks half a pint of strong decoction of devil's bit. (Tried.)

Or, use the diet drink as in the article Scorbutic Sores. I have known this cure one whose breast was as full of holes as a honey-comb.

Or, set a quart of honey by the fire to melt. When it is cold strew into it a pound and a half of quick-lime beat very fine, and sifted through a hair sieve. Stir this about till it boils up of itself into a hard lump. Beat it when cold

---

*It commonly appears first by the thickness of the lips, or a stubborn humor in the eyes, then come hard swellings in the neck chiefly, then running sores.

very fine, and sift it as before. Take of this as much as lies on a shilling, in a glass of water, every morning an hour before breakfast, at four in the afternoon, and at going to bed

Or, make a leaf of dr.ed burdock into a pint of tea ; take half a pint twice a day for four months. I have known this cure hundreds.

The best purge for the king's evil is tincture of jalap, which is made thus : Jalap, in powder, two ounces ; Geneva, or proof spirits, one pint. Let them infuse seven days. A tea-spoonful or two is sufficient for a child ten years old, in a morning, fasting ; and repeated once or twice a week, so as to keep the stomach and bowels clean, will frequently cure the king's evil. But all violent purges, when repeated too often, are pernicious.*

143. LAMENESS FROM A FIXED CONTRACTION OF THE PARTS.

Beat the yolk of a new laid egg very thin, and, by a spoonful at a time, add and beat up with it three ounces of water. Rub this gently into the parts for a few minutes three or four times a day.

144. LEGS INFLAMED.

Apply Fuller's earth spread on brown paper. It seldom fails.

---

*The tincture of jalap must be taken in any agreeable liquid.

Or, bruised or boiled turnips. Purges in most cases are absolutely necessary.

145. LEGS, SORE AND RUNNINGS.

Wash them in brandy, and apply elder leaves, changing twice a day. This will dry up all the sores, though the legs were like a honey-comb. (Tried.)

But take also a purge or two every week.

Or, poultice them with spoiled apples. (Tried.)

146. LEPROSY.*

Use the cold bath.

Or, wash in the sea often and long.

Or, mix well an ounce of pomatum, a drachm of powdered brimstone, and half an ounce of sal prunello, and anoint the parts so long as there is need.

Or, add a pint of juice of houseleek, and half a pint of verjuice, to a pint and a half of whey. Drink this, in twenty-four hours. It often cures the quinsy, and swellings on the joints.

Or, drink half a pint of celery whey, morning and evening. This has cured in a most desperate case.

Or, drink for a month a decoction of burdock leaves, morning and evening. [Tried.]

---

*In this disease the skin in many parts is covered with rough, whitish, scaly pustules, and if these are rubbed off, there remains a kind of scaly scurf.

147. LETHARGY.

Snuff strong vinegar up the nose.

Or, half a pint of decoction of water cresses, morning and evening. (Tried.)

148. LICE, [TO KILL.]*

Sprinkle Spanish snuff over the head.
Or, wash it with a decoction of amaranth.†

149. FOR ONE SEEMINGLY KILLED WITH LIGHTNING, A DAMP, OR SUFFOCATED.

Plunge him immediately into cold water.

Or, blow strongly with a bellows down his throat. This may recover a person seemingly drowned. It is still better if a strong man blows into his mouth. See the directions published by the Humane Society.

150. LUES VENEREA.

Take a little quicksilver every morning, and a spoonful of aqua sulphurata in a glass of water, at five in the afternoon. I have known a person cured by this when supposed

---

*Should be used with care.
†Prince-feather.

to be at the point of death, who had been infected by a foul nurse before she was a year old.

I INSERT this for the sake of such innocent sufferers.

151. LUNACY.

Give decoction of agrimony four times a day.

Or, rub the head several times a day with vinegar in which ground-ivy leaves have been infused.

Or, take daily an ounce of distilled vinegar.

Or, boil the juice of ground-ivy with sweet oil and white wine into an ointment. Shave the head, anoint it therewith, and chafe it every other day for three weeks. Bruise also the leaves and bind them on the head, and give three spoonsful of the juice, warm, every morning.

THIS generally cures melancholy.

The juice alone taken twice a day will cure.

Or, be electrified. (Tried.)

152. RAGING MADNESS.*

Apply to the head cloths dipped in cold water.

Or, set the patient with his head under a great water-fall

---

* It is a sure rule that all madmen are cowards, and may be conquered by binding only, without beating.—*Dr. Mead.* He also observes, that blistering the head does more harm than good. Keep the head close shaved, and frequently wash it with vinegar.

† If this is really a nervous disorder, what wonder if it should be cured by cold bathing.

as long as his strength will bear, or pour cold water on his head out of a tea-kettle.

Or, let him eat nothing but apples for a month.

Or, nothing but bread and milk. (Tried.)

153. BITE OF A MAD DOG.

Plunge into cold water daily for twenty days, and keep as long under it as possible. This has cured even after the hydrophobia was begun. †

Or, mix ashes of trefoil with hog's lard, and anoint the part as soon as possible. Repeat it twice or thrice at six hours distance. This has cured many, and particularly a dog bit on the nose by a mad dog.

Or, mix a pound of salt with a quart of water. Squeeze, bathe, and wash the wound with this for an hour; then bind some salt upon it for twelve hours.

N. B. The author of this receipt was bit six times by mad dogs, and always cured himself by this means.

Or, mix powdered liver-wort, four drachms: black pepper, two drachms. Divide this into four parts, and take one in warm milk for four mornings, fasting. Dr. Mead affirms he never knew this fail: but it has sometimes failed.

Or, take two or three spoonsful of the juice of rib-wort, morning and evening, as soon as possible after the bite. Repeat this for two or three changes of the moon. It has not been known to fail.

IMMEDIATELY consult an honest physician.

154. THE MEASLES.†

Drink only thin water-gruel, or milk and water the more the better; or toast and water.

If the cough be very troublesome, take frequently a spoonful of barley water sweetened with oil of sweet almonds newly drawn, mixed with syrup of maiden hair.

After the measles, take three or four purges, and for some weeks take care of catching cold; use light diet, and drink barley water instead of malt drink.

155. MENSES OBSTRUCTED

Be electrified. (Tried).

Or, take half a pint of strong decoction of pennyroyal every night at going to bed.

Or, boil five large heads of hemp in a pint of water to half. Strain it and drink it going to bed, two or three nights. It seldom fails. (Tried.)

Or, take six or twelve grains of calomel, in a pill, for two or three nights. taking care not to catch cold. It vomits and purges. (Tried.)

Or, pour twelve ounces of rectified spirits of wine on four ounces of roots of black hellebore, and let it stand in a

---

*This distemper is always preceded by a violent cough, often fourteen days before the red spots come out.

warm place twenty-four hours. Pour it off, and take from thirty to forty drops in any liquid, fasting.

It is good likewise in the green sickness, in all hypochondriacalcases, and in obstinate madness.

Or, burn a little sulphur of antimony on a chafing dish of coals, and receive the smoke by a funnel. In a few minutes it will take effect.

LET any of these medicines be used at the regular time, as near as can be judged.

156. MENSES NIMII.

Drink nothing but cold water with a spoonful of fine flour stirred in it. At that time drink a glass of the coldest water you can get, and apply a thick cloth dipped in cold water.

Or, put the feet into cold water.

Or, apply a sponge dipped in red wine and vinegar.

Or, bleed in the arm. Stop the orfice often with your finger, and then let it bleed again.

Or, boil four or five leaves of the red hollyhock in a pint of milk, with a small qantity of sugar. Drink this in the morning: if the person can afford it, she may add a teaspoonful of Balm of Gilead. This does not often fail.

* Or, reduce to a fine powder, half an ounce of alum with a quarter of an ounce of dragon's blood. In a violent case take a quarter of a drachm every half hour. It scarcely ever fails to stop the flux, before half an ounce is taken.

This also cures the whites.

157. TO RESOLVE COAGULATED MILK.

Cover the woman with a table cloth, and hold a pan of hot water just under her breast, then stroke it three or four minutes. Do this twice a day till it is cured.

158. TO INCREASE MILK.

Drink a pint of water going to bed.
Or, drink largely of pottage made with lentils.

159. TO MAKE MILK AGREE WITH THE STOMACH.

If it lie heavy, put a little salt in it ; if it curdle, sugar. For billious persons mix it with water.

160. MORTIFICATION, (TO STOP.)

Apply a poultice of flour, honey and water, with a little yeast.

161. NERVOUS DISORDERS.

When the nerves perform the office too languidly a good air is the first requisite. The patient should rise early, and, as soon as the dew is off the ground, walk ; let his breakfast be mother of thyme tea, gathered in June, using half as much as we do of common tea. Or, the common garden thyme, if the former cannot be procured. When the nerves are too sensible, let the person breathe a proper air. Let him eat veal, chickens, or mutton. Vegetables should be eat sparingly; the most innocent is the French bean, and the best root the turnip. Wine should be avoided carefully; so should all sauces. Sometimes he may breakfast upon a quarter of an ounce of valerian root infused in hot water, to which he may add cream and sugar. Tea is not proper. When the person finds an uncommon oppression, let him take a large spoonful of tincture valerian root.

This tincture is made thus: Cut in pieces six ounces of valerian root, gathered in June and fresh dried. Bruise it by but a few strokes in a mortar, that the pieces may be split, but it should not be beat into powder: put this into a quart of strong white wine; cork the bottle and let it stand three weeks, shaking it every day; then press it out and filter the tincture through paper.

N. B. The true wild valerian has no bad smell; if it has, cats have urined upon it, which they will do if they can come at it.

But I AM FIRMLY PERSUADED there is no remedy in nature for nervous disorders of every kind, comparable to the proper and constant use of the electrical machine.

162 NETTLE RASH.*

Rub the parts strongly with parsley. Internals profit nothing.

---

*A slight fever attended itching, smarting, and eruption, over the body

163. OLD AGE.

Take tar-water morning and evening. (Tried.)
Or, decoction of nettles; either of these will probably renew their strength for some years.
Or, be electrified daily.
Or, chew cinnamon daily.
Or, chew cinnamon daily, and swallow your spittle.

164. AN OLD STUBBORN PAIN IN THE BACK.

Steep root of water-fern, in water, till the water becomes thick and clammy; then rub the parts therewith morning and evening.
Or, apply a plaster, and take, daily, balsam of copaiba.
Or, apply garlic and hog's lard to the feet. (Tried.)

165. THE PALSY.*

Be electrified, daily, for three months. from the places wherein the nerves spring. which are brought to the paralytic part. If the parts beneath the head are affected, the fault is in the spinal marrow: if half the body, half the marrow is touched.

---

* A palsy is the loss of motion or feeling, or both, in any particular part of the body.

A PALSEY may be cured in spring or summer, but rarely in winter.

Or, use the cold bath if you are under fifty, rubbing and sweating after it.

Or, shred white onions, and bake them gently in au earthen pot till they are soft; spread a thick plaster of this and apply it to the benumbed part, all over the side if need be. I have known this cure a person seventy-five years old.

Or, take tar-water morning and evening.

THIS helps all nervous disorders.

Or, take a tea-spoonful of powdered sage lying down in bed.

166. PALSY OF THE HANDS.

Wash them often in decoction of sage as hot as you can bear.

Or, boil a handful of elder-leaves, and two or three spoonsful of mustard-seed in a quart of water. Wash often in this as hot as may be.

167. PALSY OF THE MOUTH.

After purging well, chew mustard-seed often.

Or, gargle with juice of wood sage.

168. PALSY FROM WORKING WITH WHITE LEAD OR VERDIGRISE.

Use warm baths and a milk diet.

169. PALPITATION OR BEATING OF THE HEART.

Drink a pint of cold water.
Or, apply outwardly a rag dipped in vinegar.
Or, be electrified. (Tried.)
Or, take a decoction of mother wort every night.

170. THE PILES, (TO PREVENT.)

Wash the parts daily with cold water.

171. THE PILES, (TO CURE.)

Apply warm treacle.
Or, a tobacco-leaf steeped in water twenty-four hours.
Or, a poultice of boiled brook-lime. It seldom fails.
Or, a bruised onion skinned, or roasted in ashes. It perfectly cures dry piles.
Or, varnish. It perfectly cures both the blind and bleeding piles. (Tried).
Or, fumigate with vinegar. wherein red hot flints have been quenched. This softens even scirrhus tumors.

172. THE INWARD PILES.

Swallow a pill of pitch fasting. One pill usually cures the bleeding piles.

Or, take twice a day, as much as lies on a shilling, of the thin skin of walnuts, powdered.

173. VIOLENT BLEEDING PILES.

Lightly boil the juice of nettles with a little sugar; take two ounces. It seldom needs repeating.

174. THE ·PLAGUE, (TO PREVENT.)

Eat marigold flowers, daily, as a salad, with oil and vinegar.

Or, infuse rue, sage, mint, rosemary, wormwood, of each a handful, into two quarts of the sharpest vinegar, over warm embers for eight days: then strain it through a funnel, and add half an ounce of camphire dissolved in three ounces of rectified spitits of wine. With this wash the loins, face, and mouth, and snuff a little up the nose when you go abroad. Smell to a sponge dipped therein when you approach infected persons or places.

275. THE PLAGUE, [TO CURE.]

Cold water alone, drank largely, has cured it.

Or, an ounce or two of the juice of marigolds.

Or, after bleeding fifteen or sixteen ounces, drink very largely of water sharpened with spirit of vitriol.—*Dr. Dover.*

Or, a draught of brine as soon as seized: sweat in bed; take no other drink for some hours.

Or, use lemon juice largely in every thing.

176. THE PLEURISY.

Take half a drachm of soot.

Or, take out the core of an apple, fill it up with white frankincense, stop it close with the piece you cut out, and roast it in ashes. Mash and eat it. [I.]

Or, a glass of tar-water warm every half hour.

Or, of decoction of nettles, and apply the boiled herb hot as a poultice. I never knew it fail.

Or, a plaster of flour of brimstone and white of an egg. (Tried.) This seldom fails.

In disordes of this kind Dr. Huxham advises: "Sip almost continually thin whey, barley-water, or hyssop tea, sharpened with vinegar and water or lemon-juice. If the spitting stop suddenly, take a gentle vomit. Likewise camphorated vinegar, with syrup of elder or raspberries, is good." To appease the cough, take often, a little at a time, of roasted apples, of strawberries, raspberries, or currants

177. TO ONE POISONED.

Give one or two grains of distilled verdigrise. It vomits in an instant.

Let one poisoned by arsenic dissolve a quarter of an ounce

of salt of tartar in a pint of water, and drink every quarter of an hour as much as he can, till he is well.

Let one poisoned by opium take thirty drops of elixir of vitriol in cold water, every quarter of an hour, till the drowsiness or wildness ceases.

Or, a spoonful of lemon-juice.

Let one poisoned by mercury sublimate dissolve an ounce of salt of tartar in a gallon of water, and drink largely of it.

THIS will entirely destroy the force of the poison if it be used soon.

Nothing cures the African poison but a decoction of the roots of the sensitive plant

178. A POLYPUS IN THE NOSE.

Powder a lump of alum and snuff it up frequently: then dissolve powdered alum in brandy, dip lint therein, and apply it at going to bed.

179. A PRICK OR CUT THAT FESTERS.

Apply turpentine.

180. PTYALISM, OR CONTINUAL SPITTING.

A very violent and stubborn disorder of this kind was cured by chewing perpetually a little dry bread, and swallowing it with the spittle.

181. AN EASY PURGE.

Drink a pint of warmish water fasting, walking after it.

Or, a soft egg with a tea-spoonful of salt.

Or, infuse from half a drachm to two drachms of damask rose leaves, dried, in half a pint of warm water for twelve hours, and take it.

Or, infuse three drachms of senna, and a scruple of salt of tartar, in half a pint of river water for twelve hours; then strain and take it in the morning.

Wild ash is a plant of the very same nature with senna. Its leaves taken in double the quantity purge full as well and does not gripe as senna does.

The wild ash is called, in the north of England, round-tree, quick-beam, or wigan-tree. The leaves should be gathered when the tree is in flower.

182. A STRONGER PURGE.

Drink half a pint of strong decoction of dock-root.

Or, the jalap powder in treacle, or any liquid in the morning fasting.

Or, the jalap powder may be made into pills.

Or, a table-spoonful of tincture of jalap in a morning, fasting, in a cup of cold camomile tea.

183. THE QUINSY*

Apply a large white bread toast half an inch thick, dip-

---
*Fever attended with difficulty of swallowing, and often of breathing.

ped in brandy, to the crown of the head till it dries.

Or, swallow slowly white rose-water mixed with syrup of mulberries. (Tried.)

Or, juice or jelly of black currants, or the decoction of the leaves or bark.

Or, draw in as hot as you can bear, for ten or twelve minutes together, the fumes of red-rose leaves, or camomile flowers, boiled in water and vinegar, or of a decoction of bruised hemp-seed.

This speedily cures the sore throat, peripneumony, and inflammation of the uvula.

184. A QUINSY OF THE BREAST.*

Take from eight to twenty drops of laudanum lying down in bed. This helps.

Or, make an issue in the thigh. This cures.

185. THE RHEUMATISM.†

To prevent, wear washed wool under the feet.
To cure, use the cold bath with rubbing and sweating.
Or, apply warm steams.
Or, rub in warm treacle, and apply to the part brown pa-

---

* This is known by a sudden unaccountable pain and difficulty of breathing seizing a person in the night, or any violent motion.

† Rheumatical pains are generally most violent as soon as you are warm in bed : but there is a cold rheumatism which is most painful when the part is cold. Constant rubbing will cure this.

per smeared therewith ; change it in twelve hours. [Tried.]

Or, drink half a pint of tar-water morning and evening.

Or, steep six or seven cloves of garlic in half a pint of white wine; drink it lying down. It sweats and frequently cures at once.

Or, mix flour of brimstone with honey, in equal quantities, take three teaspoonsful at night, two in the morning, and one afterwards, morning and evening, till cured. This succeeds oftener than any remedy I have found.

Or, live on new milk whey and white bread for fourteen days. This has cured one in a desperate case.

Or, pound the green stalks of English rhubarb in May or June, with an equal quantity of lump sugar. Take the quantity of a nutmeg of this three or four times a day. This seldom fails.

In a stubborn rheumatism, let your diet be barley-gruel, with currants, roasted apples, fresh whey, and light pudding.

186. TO RESTORE THE STRENGTH AFTER A RHEUMATISM.

Make a strong broth of cow-heels, and wash the parts with it warm twice a day. It has restored one who was quite a cripple, having no strength left either in his leg thigh, or loins.

Or, mix gum guaiacum in powder, with honey and treacle: take two or three tea-spoonsful [or as much as you can bear without purging], twice or thrice a day. This is the best medicine I have met with for the chronic rheumatism.

Or, dissolve one ounce of gum guaiacum in three ounces of spirits of wine: take sixty or eighty drops on loaf sugar, two or three times a day. This is Dr. Hill's Essence of Bardana.

Or, drop thirty drops of volatile tincture of guaiacum on a lump of sugar. and take this in a glass of water every four hours. It usually cures in a day. (Tried.)

187. RICKETS, (TO CURE OR PREVENT.)

Wash the child every morning in cold water.

188. RING WORMS.*

Apply rotten apples or pounded garlic.
Or, rub them with juice of house-leek.
Or, wash teem with Hungary water camphorated.
Or, twice a day with oil of sweet almonds and oil of tartar mixed.

189. RUNNING AT THE NOSE.

Snuff up a tea-spoonful of spirits of hartshorn

190. A RUPTURE.

Foment with hot aqua vitæ for two hours.

---

* Vulgarly called tetters.

Or, take agrimony, spleen-wort, Solomon's seal, strawberry roots, a handful of each; pick and wash them well; stamp and boil them two hours in two quarts of white wine in a vessel close stopped: strain and drink a large glass of this every morning, and an hour after drink another. It commonly cures in a fortnight. A good truss in the meantime is of great use, and perhaps the only thing to be depended on.

"I place," says Dr. Riviere, "a broad plank sloping from the side of the bed to the floor. On this I lay the patient upon pillows, with his head downward. Then I foment the part for half an hour with cloths, four times doubled, steeped in cold water, gently touching it with my fingers. Afterwards I bind on it, many times doubled, a cloth shaped like a triangle, wet in cold water. The gut is generally restored to its place in a few hours. If not, I repeat the operation twice a day, and in two or three days the disease is cured."

191. A RUPTURE IN CHILDREN.

Boil a spoonful of egg-shells dried in an oven, and powdered, in three quarters of a pint of milk. Feed the child constantly with bread boiled in this milk.

192. A WINDY RUPTURE.

A poultice of the entrils of a cow on leather, strewing some cummin seeds on it, and apply it hot. When cold put on a new one. It commonly cures a child (keeping in bed) in two days.

193. A SCALD HEAD.

Anoint it with Barbadoes tar.

Or, apply daily white wine vinegar. (Tried.)

If wood soot is mixed with fresh butter into an ointment, and the head anointed with it every day it will generally cure it at the beginning; but when it is become very bad, a a plaster should be made of gall, dried to the consistence of salve, and spread upon linen. This should be applied all over the parts affected, and continued on four or five days; then it should be taken off and the head dressed with soot ointment as before.

After the cure, give two or three gentle purges.

If a proper regard was paid to cleanliness in the head and apparel of children, the scald head would seldom be seen.

194. THE SCIATICA.*

Is certainly cured by a purge taken in a few hours after it begins.

Or, use cold bathing and sweat, together with the flesh brush twice a day.

Or, boil nettles till soft: foment with the liquor, then apply the herb as a poultice. I have known this cure a sciatica of forty-five years standing.

Or, a mud made made of powdered pit coal and warm water. This frequently cures sores, weakness of the limbs,

---

*The sciatica is a violent pain in the hip, chiefly in the joints of the thigh bone.

most disorders of the legs, and swellings of the elbow joint, though accompanied with a fistula arising from a caries of the bone.

195. INFLAMMTION OR SWELLING OF THE SCROTUM.

Wash it thrice a day with a strong decoction of agrimony

196. A SCORBUTIC ATROPHY.*

Use cold bathing. Which also cures all scorbutic pains

197. SCORBUTIC GUMS.

Wash them daily with the decoction of the Peruvian bark, adding a little tincture of rosemary, with a solution of myrrh.

198. SCORBUTIC SORES.

A diet drink : Put half a pound of fresh-shaved lignum guaiacum [called by the blockmakers, lignum vitæ], and half an ounce of senna, into an earthen pot that holds six quarts. Add five quarts of soft water and lute the pot close

---

* Such a degree of the scurvy as causes the flesh to waste away like a consumption.

Set this in a kettle of cold water, and put it over a fire till it has boiled three hours. Let it stand in the kettle till cold. When it has stood one night, drink daily half a pint in new milk warm, in the morning, fasting, and at four in the afternoon unless it purges too much, if so, take less. Wash with a little of it. In three months all the sores will be dried up (Tried).

199. SCURVY.†

Live on turnips for a month.

Or, take tar water, morning and evening, for three months.

Or, three spoonsful of nettle juice every morning. (Tried.)

*Or, decoction of burdock. Boil three ounces of the dried root in two quarts of water to three pints. Take a half pint daily. A decoction of the leaves (boiling one leaf four minutes in a quart of water,) has the same effect.

Or, take a cupful of the juice of goose grass in a morning fasting, for a month; it is frequently called hariff, or cleavers, I have known many persons cured by it.

Or, pound into a pulp, of Seville orangers, sliced, rind and all, and powder sugar, equal quantities. Take a tea-spoonful three or four times a day. (Tried.)

Or, squeeze the juice of half a Seville orang into a pint of

---

† The scurvy is known by heaviness of the body, weariness, rottenness of the gums, and yellow, lead, or violet colored spots on the legs or arms.

N. B. A scurvy attended with costiveness (which is the most common) is termed a hot scurvy, one attended with looseness, a cold scurvy.

milk over the fire. Sweeten the whey with loaf sugar, and drink it every morning new milk warm. To make any whey, milk should be skimmed after it is boiled.

Or, pour three quarts of boiling water on a quart of ground malt; stir them well, and let the mixture stand close covered for four hours; strain it off and use this as common drink: in hot weather, brew this fresh every day. It will hardly fail.

Or, take morning and evening a spoonful or two of lemon juice and sugar. It is a precious remedy, and well tried.— *Dr. Macbridge.*

Water and and garden cresses, mustard, and juice of scurvy grass help in a cold scurvy.

When there is a continual salt taste in the mouth, take a pint of lime-water morning and evening.

200. A BROKEN SHIN.

Bind a dry oak leaf upon it.

Or, put on a bit of white paper moistened with spittle. It will stay on till the place is well. (Tried.) This cures a cut.

201. SHINGLES.*

Drink sea water every morning for a week: towards the close bathe also.

Or, apply pounded garlic.

---

* A kind of ring worm which encircles the body like a belt of a hand's breadth.

202. SICKISHNESS IN THE MORNING.

Eat nothing after six in the evening.
Or, drink half a pint of water impregnated with fixed air.

203. SINEWS SHRUNK.

Rub the part every morning with fasting spittle.
Or, beat the yolk of a new-laid egg, mix it with a spoonful of water, and rub the part with it before the fire three or four times a day.

204. SKIN RUBBED OFF.

Apply pounded all-heal. It seldom needs repeating.
Or, a bit of white paper with spittle.

205. SMALL POX.

Drink largely of toast and water.
Or, let your whole food be milk and water, mixed with a little white bread.
Or, milk and apples.
Take good care to have free, pure and cool air. Therefore open the casement every day; only do not let it chill the patient
If they strike in, and convulsions follow, drink a pint of cold water immediately. This instantly stops the convulsion, and drives out the poek. (Tried.)

There may be pustules a second time, coming out and ripening like the small pox, but it is barely a cutaneous disorder.

"In violent cases, bleed in the foot ; bathe the legs in warm water twice or thrice a day, before and at the eruption, and apply boiled turnips to the feet. Never keep the head too hot.

"In very low depressed cases wine may be given, and if the pustules lie buried in the skin, a gentle vomit ; in many cases a gentle purge of manna, cream of tartar or rhubarb.

"In the Crude Ichorose small pox, a dish of coffee now and then, with a little thick milk in it, has often quieted the vexatious cough.

"After the incrustation is formed, change the sick, but let it be with very dry warm linen.—*Dr. Huxham.*

206. A LONG RUNNING SORE IN THE BACK.

Was entirely cured by eating betony in everything.

Or, take every morning two or three spoonsful of nettle juice, and apply nettles bruised in a mortar to the part. This cures any old sore or ulcer. (I.)

207. A SORE LEG.

Bind a diculum plaster, an inch broad, round the leg, just above the sore, and foment it morning and evening with hot water.

Any sore is healed by a plaster of mutton-suet, even though it fester or breed proud flesh.

208. A SORE MOUTH.

Apply the white of an egg beat up with loaf sugar.
Or, gargle with the juice of cinquefoil.
Or, boil together a pound of treacle, three yolks of eggs, an ounce of bole amoniac, and the quantity of a nutmeg of alum a quarter of an hour. Apply this to the sore part, or to an aching tooth. (Tried.)

209. A SORE THROAT.

Take a pint of cold water lying down in bed. (Tried.)
Or, apply a chin-stay of roasted figs.
Or, a flannel sprinkled with spirits of hartshorn to the throat, rubbing Hungary water on the top of the head. (Tried.)
Or, snuff a little honey up the nose.
An old sore throat was cured by living wholly on apples and apple-water.

210. AN INFLAMED SORE THROAT.

Lay nitre and loaf sugar, mixed, on the tongue.

211. A PUTRID SORE THROAT.

Lay on the tongue a lump of sugar dipped in brandy.

212. A SPRAIN.

Hold the part in very cold water for two hours. (Tried.)
Or, apply cloths dipped therein, four times doubled, for two hours, changing them as they grow warm.
Or, bathe it in good crab verjuice.
Or, boil bran in wine vinegar to a poultice. Apply this warm and renew it once in twelve hours.
Weakness remaining afrer a sprain, is cured by fomenting the part daily with beef-brine.
Suppose the ancle sprained : 1st. Foment it with warm vinegar four or five times every four hours. 2nd. Stand, if you can, three or four minutes at a time on both your feet, and frequently move the sprained foot. Sometimes also while sitting with your foot on a low stool, move it to and fro. 3rd. Let it be gently rubbed with a warm hand at least thrice a day. 4. Two hours after every application of the vinegar, let it be just wetted with spirits of wine and then gently rubbed.

213. A VENOMOUS STING.

Apply the juice of honey-suckle leaves.
Or, a poultice of bruised plantain and honey.
Or, take inwardly, one drachm of black currant leaves powdered. It is an excellent counter-poison.

214. THE STING OF A BEE.

Apply honey.

Or, mix a little turpentine with flour in the yolk of an egg, and apply it as a plaster. This cures in a desperate case.

215. THE STING OF A NETTLE.

Rub the part with the juice of nettles.

216. THE STING OF A WASP.

Rub the part with the bruised leaves of house-leek, water-cresses or rue.

Or, apply treacle or sweet oil.

Or, bruised onions or garlic.

217. STING OF BEE OR WASP IN THE EYE.

Apply carduus bruised with the white of an egg, renew it if it grows dry.

218. STING IN THE GULLET.

Beat well together, with a spoon, some honey and sweet oil with a little vinegar; swallow a spoonful every minute till ease is procured.

219. A STITCH IN THE SIDE.

Apply treacle spread on hot toast. (Tried.)

220. ACCIDENTAL SICKISHNESS, OR PAIN IN THE STOMACH.

Vomit with a quart of warm water. Do this twice or thrice, omiting a day between.

221. PAIN IN THE STOMACH FROM BAD DIGESTION.

Take fasting, or in the fit, half a pint of camomile tea. Do this five or six mornings.

Or, drink the juice of half a large lemon, or sweet orange, immediately after dinner every day.

Or, in the fit a glass of vinegar.

Or, take two or three tea spoonsful of stomachic tincture in a glass of water, thrice a day.

THE tincture is made thus: Gentian root, sliced, one ounce; orange peel, dried, half an ounce; cochineal, fifteen grains; proof brandy one pint: in three or four days it is fit for use. This is useful in all disorders that arise from a relaxed stomach.

222. CHOLERIC PAINS IN THE STOMACH.

Take half a pint of decoction of ground-ivy, with a teaspoonful of the powder of it, five or six mornings. (I.)

223. COLDNESS IN THE STOMACH.

Take a spoonful of the syrup of the juice of carduus benedictus, fasting, for three or four mornings. (I.)

Or, chew a leaf of carduus every morning, and swallow the spittle. (Tried).

224. PAIN IN THE STOMACH, WITH COLDNESS AND WIND.

Swallow five or six corns of pepper for six or seven mornings. (Tried.)

225. STONE, (TO PREVENT.)

Eat a small crust of bread every morning. (Tried).

Or, drink a pint of warm water, daily, just before dinner. After discharging one stone, this will prevent the generating of another. Stoop down and raise up again. If you feel pain as if cut through the middle, the pain is not from the stone, but rheumatism. Beware of costiveness. Use no violent diuretics. Mead is a proper drink.

Or, slice a large onion, pour half a pint of warm water upon it. After it has stood twelve hours, drink the water. Do this every morning till you are well.

226. IN A RAGING FIT.

* Beat onions into a pulp and apply them as a poultice to the back, or to the groin. It gives speedy ease in the most racking pain. (Tried.)

227. STONE, (TO EASE OR CURE.)

Boil half a pound of parsnips in a quart of water. Drink a glass of this morning and evening, and use no other drink all the day. It usually cures in six weeks.

Or, take morning and evening a tea-spoonful of onions, calcined in a fire shovel into white ashes, in white wine. An ounce will often dissolve the stone.

Or, take a tea-spoonful of violet seed, powdered, morning and evening. It both wastes the stone and brings it away.

Or, drink largely of water impregnated with fixed air.

Those who have not a convenient apparatus, may substitute the following method : Dissolve sixteen grains of salt of tartar in six spoonsful of water, to which add as much water aciduated with oil of vitriol as will neutralize the salt. They are to be gradually mixed with each other, so as to prevent the efferescence or dissipation of the fixed air as much as possible.

228. STONE IN THE KIDNEYS.

Use the cold bath.
Or, drink half a pint of water every morning.
Or, boil an ounce of common thistle-root, and four

drachms of liquorice, in a pint of water. Drink of it every morning.

229. STOPPAGE IN THE KIDNEYS.

Take decoction, or juice, or syrup of ground-ivy, morning and evening.

Or, half a pint of tar-water.

Or, twelve grains of salt of amber in a little water.

230. THE STRANGUARY.

Sit over the steam of warm water.

Or, drink largely of decoction of turnips, sweetened with clarified honey.

* Or, of warm lemonade. (Tried.)

Or, dissolve half an ounce of saltpeter in a quart of water. Drink a glass of it every hour.

231. SUNBURN, (SMARTING.)

Wash the face with sage tea.

232. A FRESH SURFEIT.

Take about the size of a nutmeg of the green tops of wormwood.

233. TO STOP PROFUSE SWEATING.

Drink largely of cold water.

234. TO PREVENT IT.

Mix an ounce of tincture of Peruvian bark with half an ounce of spirit of vitriol. Use a tea-spoonful morning and evening in a glass of water.

235. TO CURE NIGHT SWEATS.

Drink a gill of warm milk at lying down.

236. SWELLED GLANDS IN THE NECK.

Take sea-water every other day.

237. INDOLENT SWELLINGS

Are often cured by warm sweats.

238. SOFT AND FLABBY SWELLINGS.

Pump cold water on them daily.
Or, use constant friction.
Or, proper bandages.

239. WHITE SWELLING IN THE JOINTS.

Hold the part half an hour every morning under a pump

or cock. This cures all pains in the joints. It seldom fails. (Tried.)

Or, pour on it daily a stream of warm water.

Or, a stream of cold water one day, and warm the next, and so on, by turns.

Use these remedies at first, if possible. It is likewise proper to intermix gentle purges to prevent a relapse.

Or, boiled nettles.

240. TO DISSOLVE WHITE OR HARD SWELLINGS.

Take white roses, elder flowers, leaves of fox-gloves, and and of St. John's wort, a handful of each; mix with hog's lard and make an ointment.

Or, hold them morning and evening in the steam of vinegar poured on red hot flints.

241. TO FASTEN THE TEETH.

Put powdered alum, the quantity of a nutmeg, in a quart of spring water for twenty-four hours: then strain the water and gargle with it.

Or, gargle often with phyllera leaves boiled with a little alum in forge water.

242. TO CLEAN THE TEETH.

Rub them with ashes of burned bread.

243. TO PREVENT THE TOOTH-ACHE.

Wash the mouth with cold water every morning, and rince them after every meal.

Or, rub the teeth often with tobacco ashes.

244. TO CURE THE TOOTH-ACHE.

Be electrified through the teeth. (Tried.)

Or, apply to the aching tooth an artificial magnet.

Or, rub the cheek a quarter of an hour.

Or, lay roasted parings of turnips, as hot as may be, behind the ear.

Or, put a leaf of betony, up the nose.

Or, lay bruised or boiled nettles to the cheek. (Tried.)

Or, lay a clove of garlic on the tooth.

Or, hold a slice of apple lightly boiled between the teeth. (Tried.)

Or, keep the feet in warm water, and rub them well with bran just before bed time. (Tried.)

THE first twenty teeth generally last till the sixth or seventh year; after that, till the fourteenth or fifteenth year; they fall out one by one, and are succeeded by others.

The shedding of teeth is wisely intended, and brought about in a singular manner. Their hardness will not admit of distension like other parts of the body. Hence, after an enlargement of the jaw-bone, the original teeth are no longer able to fill up the cavities of it. They must stand unsupported by each other, and leave spaces between them.

Under the first teeth, therefore, is placed a new set, which, by constant pressing upon their roots, rob them of their nourishment, and finally push them out of their sockets.

245. TOOTH-ACHE FROM COLD AIR.

Keep the mouth full of warm water.

246. TEETH SET ON EDGE.

Rub the tops of the teeth with a dry towel.
There is no such thing as worms in the teeth. Children's using coral is always useless, often hurtful.

" Forcing the teeth into order is always dangerous. Filling is generally hurtful.

"All rough and cutting powders destroy the teeth, so do all common tinctures.

" Sweet meats are apt to hurt the teeth, if the mouth be not rinsed after them. Cracking nuts often breaks off the enamel; so does biting thread in two.

" Constant use of tooth-picks is a bad practice, constant smoking of tobacco destroys many good sets of teeth."–*Dr. Beardmore.*

247. EXTREME THIRST, [WITHOUT A FEVER.]

Drink spring water in which a little sal prunello is dissolved.

248. PAIN IN THE TESTICLES.

Apply pellitory of the wall beaten up into a poultice changing morning and evening.

249. TESTICLES INFLAMED.

Boil bean flower in three parts of water, one part vinegar.

250. TO DRAW OUT THORNS, SPLINTERS, AND BONES.

Apply nettle-roots and salt.
Or, turpentine spread on leather.

251. THRUSH.*

Mix juice of calendine with honey to the thickness of cream. Infuse a little powdered saffron, let this simmer a while and scum it; apply it where needed with a feather. At the same time give eight or ten grains of rhubarb; to grown persons twenty.

Or, take an ounce of clarified honey, having scummed off all the dross from it, put in a drachm of roche-alum finely powdered, and stir them well together. Let the child's mouth be rubbed well with this five or six times a day, with

---

* Little white ulcers in the mouth.

a bit of rag tied upon the end of a stick ; and though it be the thorough thrush, it will cure it in a few days. I never knew it fail.

Or, burn scarlet cloth to ashes and blow them into the mouth. This seldom fails.

252. TONSILS SWELLED.

Wash them well with lavender-water.

253. TORPOR OR NUMBNESS OF THE LIMBS.

Use the cold bath with rubbing and sweating.

254. TWISTING OF THE GUTS.

See the Second Part of this book for the remedy, under the head of Billious Colic.

255. TYMPANY OR WINDY DROPSY.

Use the cold bath with purges intermixed.

Or, mix juice of leeks and of elder. Take two or three spoonsful of this morning and evening. (Tried.)

Or, eat a few parched peas every hour.

256. A VEIN OR SINEW CUT.

Apply the inner green rind of hazel, fresh scraped.

257. THE VERTIGO, OR SWIMMING IN THE HEAD.

Take a vomit or two.

Or, use the cold bath for a month.

Or, in a May morning about sun-rise snuff up the dew, daily, that is on mallow leaves.

Or, apply to the top of the head, shaven, a plaster of flour of brimstone and white of eggs. (Tried.)

Or, take every morning half a drachm of mustard-seed.

Or, mix one part of salt of tartar with three parts of cream of tartar; take a tea-spoonful in a glass of water, every morning, fasting. This is serviceable when the vertigo springs from acid, tough phlegm in the stomach.

258. VIGILIA, OR INABILITY TO SLEEP.

Apply to the forehead, for two hours, cloths four times doubled, dipped in cold water. I have known this applied to a lying-in woman and her life saved thereby.

Or, take a grain or two of camphire. It is both safer and surer than opium.

Assafœtida, from ten to thirty grains, likewise will in most cases have as much effect as opium. Drink no green tea in the afternoon.

259. BITE OF A VIPER OR RATTLE SNAKE.

Apply bruised garlick.

Or, rub the place immediately with common oil. Query, would not the same cure the bite of a mad dog? Would it not be worth while to make a trial on a mad dog?

260. TO PREVENT THE BITE OF A VIPER

Rub the hands with the juice of radishes.

261. AN ULCER.

Dry and powder a walnut leaf and strew it on, and lay another walnut leaf on that. (Tried.)

Or, boil walnut tree leaves in water with a little sugar. Apply a cloth dipped in this, changing it once in two days. This has done wonders.

Or, foment morning and evening with a decoction of walnut-tree leaves on. This has cured foul bones; yea, and a leprosy. (Tried.)

262. ULCER IN THE BLADDER OR KIDNEYS.

Take a decoction of agrimony thrice a day.
Or, decoction, powder, or syrup of horsetail.

263. ULCER IN THE GUM OR JAW.

Apply honey of roses sharpened with spirits of vitriol.

Or, fill the whites of eggs boiled hard and slit, with myrrh and sugar-candy, powdered. Tie them up, and hang them on sticks lying across a glass. A liquid distils, with which anoint the sores often in a day.

264. A FISTULOUS ULCER.

Apply wood-beteny, bruised, changing it daily.

265. A BLEEDING VARICOUS ULCER IN THE LEG.

Was cured by constant cold bathing.

266. A MALIGNANT ULCER.

Foment, morning and evening, with a decoction of mint; then sprinkle on it finely powdered rue.

Or, burn to ashes, but not too long, the gross stalks on which the red coleworts grow. Make a plaster with fresh butter. Change it once a day.

Or, apply a poultice of boiled parsnips. This will cure even when the bone is foul.

Or, be electrified. (Tried.)

267. AN ULCER IN THE URETHRA*

Take a clove of garlic morning and evening.

---

* The urethra is the passage of the urine.

268. AN EASY AND SAFE VOMIT.

Pour a dish of tea on twenty grains of ipecacuanha. You may sweeten it if you please. When it has stood four or five minutes, pour the tea clear off and drink it.

269. TO STOP VOMITING.

Apply a large onion, slit across the grain, to the pit of the stomach. (Tried.)

Or, take a spoonful of lemon juice and six grains of salt of tartar.

270. BLOODY URINE.

Take twice a day a pint of decoction of agrimony.
Or, of decoction of yarrow.

271. URINE BY DROPS, WITH HEAT AND PAIN.

Drink nothing but lemonade. (Tried)

Or, beat up the pulp of five or six roasted apples with a quart of water. Use it at lying down. It commonly cures before morning.

272. INVOLUNTARY URINE.

Use the cold bath.

Or, take a tea-spoonful of powdered agrimony in a little water morning and evening.

Or, a quarter of a pint of alum posset drink every night.

Or, foment with rose leaves and plantain leaves, boiled in smith's forge water; then apply plasters of alum and bole armoniac made up with oil and vinegar.

Or, apply a blister to the os-sacrum. It seldom fails.

273. SHARP URINE.

Take two spoonsful of fresh juice of ground-ivy.

274. SUPPRESSION OF URINE.

Sometimes relieved by bleeding

Or, drink largely of warm lemonade.

Or, a scruple of nitre every two hours.

Or, take a spoonful of the juice of lemon sweetened with syrup of violets.

Or, seven grains of mercurius dulcis.

275. UVULA INFLAMED.*

Gargle with a decoction of beaten hemp-seed.

Or, with a decoction of dandelion.

Or, touch it frequently with camphorated spirits of wine·

---

* This is usually called the palate of the mouth.

276. UVULA RELAXED.

Bruise the veins of a cabbage leaf and lay it hot on the crown of the head; repeat, if needed, in two hours. I never knew it fail.

Or. gargle with an infusion of mustard-seed.

277. WARTS.

Rub them daily with a radish.
Or, water in which sal amoniac is dissolved.
Or, with juice of marigold flowers: it will hardly fail.
Or, apply bruised purslain as a poultice, changing it twice a day. It cures in seven or eight days.

278. WEAKNESS IN THE ANCLES.

Hold them in cold water a quarter of an hour morning and evening.

279. A SOFT WEN.

Wrap leaves of sorrel in a wet paper, and roast them in the embers: mix it with finely sifted ashes into a poultice. Apply this warm, daily.

Dr. Reviere says: "I cured a wen as big as a large fist thus: I made an instrument of hard wood, like the stone with which the painters grind their colors on a marble.

With this I rubbed it half an hour twice a day; then I laid on a suppurating plaster very hot, which I kept on four or five days. The wen suppurated and was opened. Afterwards all the substance turned into matter, and was evacuated. Thus I have cured many since.

280. THE WHITES.

Live chastely; feed sparringly; use exercise constantly; sleep moderately, but never lying on your back.

Take eight grains of jalap every eight days. This usually cures in five weeks.

Or, first bleed; then purge thrice with twenty grains of rhubarb, and five of calomel.

Or, boil four or five leaves of the white holly-hock in a pint of milk with a little sugar, then add a tea-spoonful of balm of Gilead. Drink this every morning. It rarely fails.

Or, make Venice turpentine, flour, and fine sugar, equal quantities, into small pills. Take three or four of these morning and evening. This also cures most pains in the back.

Or, take yellow resin, powdered, one ounce; conserve of roses, half an ounce; powdered rhubarb, three drachms; syrup a sufficient quantity to make an electuary. Take a large tea-spoonful of this twice a day, in a cup of confrey-root tea.

Or, in a quarter of a pint of water, wherein three drachms of tamarinds and a drachm of lentisk-wood has been boiled. When cold, infuse senna, one drachm; coriander-seed and liquorice, a drachm and a half each. Let them stand all night. Strain the liquor in the morning, and drink it daily two hours before breakfast.

281. A WHITLOW.

Apply treacle. (Tried.)
Or, honey and flour. (Tried.)
Or, a poultice of chewed bread. Shift it once a day.
Or, a poultice of powdered pit coal and warm water.

*282. WORMS.

Use two tea-spoonsful of brandy, sweetened with loaf sugar, every morning.

Or, a spoonful of the juice of lemons.

Or, two spoonsful of nettle juice.

Or, boil four ounces of quicksilver an hour in a quart of clear water: pour it off and bottle it up. You may use the same quicksilver again and again. Use this for common drink; or at least, night and morning for a week or two. Then purge off the dead worms with fifteen or twenty grains of jalap.

Or, take two tea-spoonsful of worm seed mixed with treacle, for six mornings.

---

* A child may be known to have the worms by chilliness, paleness, hollow eyes, itching of the nose, starting in sleep, and an unusual stinking breath. Worms are never found in children that live wholly on milk.

Or, one, two, or three drachms of powdered fern-root boiled in mead. This kills both the flat and round worms. Repeat the medicine from time to time.

Or give one tea-spoonful of syrup of bear's foot at bed time, and one or two in the morning, for two or three successive days, to children between two and six years of age, regulating the dose according to the strength of the patient.

Syrup of bear's foot is made thus: Sprinkle the green leaves with vinegar, stamp and strain out the juice, and add to it a sucfficient quantity of coarse sugar. This is the most powerful medicine for long round worms.

Bruising the green leaves of bear's foot and smelling often at them, sometimes expels worms.

Or, boil half an ounce of aloes, powdered, with a few sprigs of rue, wormwood, and camomile, in a half a pint of gall, to the consistency of a plaster; spread this on thin lether and apply it to the stomach, changing it every twelve hours, for three days; then take fifteen grains of jalap, and it will bring vast quantities of worms away, some burst, and some alive. This will cure when no internal medicine avails.

283. FLAT WORMS.

Mix a tablespoonful of Norway tar in a pint of small beer; take it as you can in the morning, fasting. This brought away a tape worm thirty-six feet in length.

Or take from two to five grains of gamboge made into a bolus, in the morning, fasting, drinking after it a little weak green tea, and likewise when it begins to operate, till the worm is evacueted. The dose must be regulated according to the patient's strength; for neither this nor any other medicine given as an alterative, is of the least service in this disorder. If the head of the worm be fixed in the upper orfice of the stomach, a smart shock from the electrifying machine will probably dislodge it: then purge. To prevent —avoid drinking stagnated water.

284. WOUNDS.

If you have not an honest surgeon at hand
Apply juice or powder of yarrow—(I.)
Or, bind leaves of ground-ivy upon it.
Or, wood-betony bruised. It quickly heals even cut veins and sinews, and drawsout thorns or splinters.

Or. keep the part in cold water for an hour, keeping the wounds closed with your thumb; then bind on the thin skin of an egg-shell for days or weeks, till it falls off itself. Regard not though it prick or shoot for a time.

285 INWARD WOUNDS.

Infuse yarrow twelve hours in warm water. Use a cup of this four times a day.

286. PUTRID WOUNDS.

Wash them morning and evening with warm decoction of agrimony. If they heal too soon, and a matter gathers under-neath, apply a poultice of the leaves pounded, changing them once a day till well.

Or, apply a carrot poultice ; but if the gangrene comes on, apply a wheat poultice (after it has been by the fire till it begins to ferment) nearly cold. It will not fail.

287. WOUNDED TENDONS.

Boil comfrey roots to a thick mucilage or jelly, and apply this as a poultice, changing it twice a day.

288. TO OPEN A WOUND THAT IS CLOSED TO SOON.

Apply bruised centuary.

MEDICINES.

DAFFY'S ELIXIR.

TAKE of the best senna, guaiacum, liquorice sliced small, aniseeds, coriander seeds, and elecampane-root, of each half an ounce; raisins of the sun,* stoned, a quarter of a pound: let them all be bruised and put into a quart of the best brandy. Let it stand by the fire a few days, and then strain it.

ANOTHER RECEIPT FOR DAFFY'S ELIXIR.

Take of senna leaves, two ounces † coriander-seeds, a quarter of an ounce; proof spirit, or brandy, three pints: put all the ingredients into a bottle for four or five days, shaking it frequently; strain off the tincture, and add three ounces of powdered sugar candy. This medicine is more active than the preceding, and is calculated to remove obstructions in the bowels, in cholics and other complaints that require purging, especially when castor oil has not had the

---

* Sun cured. † Jalap one ounce.

desired effect. The dose is one, two, or three table-spoonsful, in a cup of camomile tea, or water.

TURLINGTON'S BALSAM.

Take balsams of Peru and Tolu, of each half an ounce; gum storax, in tears, and gum guaiacum, of each one ounce; gum benjamin, an ounce and a half; hepatic aloes and frankincense, of each two drachms: let the gum be bruised, and put all the ingredients into a quart of rectified spirits of wine; shake the bottle frequently, and in eight days it is fit for use.

This is indeed a most excellent medicine for man or beast, or for any fresh wound. I know none like it.

SCOTCH PILLS.

Dissolve two ounces of hepatic aloes, with a small spoonful of sweet oil and as much water, in a porringer over a small fire. When it is of a proper consistence, make it into pills with or without liquorice powder.

EMETIC TARTAR VOMIT.

Dissolve four grains of emetic tartar in half a pint of hot water; stir it about well: when it is cold it is fit for use. Take two table-spoonsful every quarter of an hour till it operates; after which no more of the vomit must be taken. Drink a small cup of gruel, or weak camomile tea after every puke to work it off. A pint or a pint and a half of gruel of tea is generally sufficient. To settle the stomach, drink a little weak brandy and water, and lie down half an hour.

One table-spoonful of the emetic tartar water, every quarter of an hour till it pukes, is sufficient for weak people; while others again require four times as much. A child of a month old may take a small tea-spoonful every quarter of an hour ; one of three months old will require two tea-spoonful, and so in proportion to their age and strength. Children require nothing to work off a vomit ; and a pint or a pint and a half of gruel or camomile tea is sufficient for adults. It is an absurd and pernicious practice to drink pint after pint of hot liquids to work off a vomit, and frequently leaves a very great relaxation of the stomach, which does not recover its tone for some months afterwards.

The design of giving a vomit in the manner above described, is in order that it may work in the gentlest manner possible. If it operates two, or three, or four times, it is sufficient. Violent vomits are often attended with dangerous consequences; whereas gentle ones may be repeated two or three times a week if necessary.

If a vomit works too violently, drink moderately of weak brandy and water, and apply a raw onion cut in two to the pit of the stomach.

The best time for taking a vomit is in a morning, fasting. But in cases where no time is to be lost, it may be taken at eleven o'clock, or in the evening.

Persons who are costive, should not venture upon a vomit till the costiveness is removed, which must be done in an hour or two's time by a clyster, or a small dose of jalap powder, or any other opening medicine.

In consumptive cases, and in the dysentery cases, ipecacuanha is the proper vomit. The emetic tartar is best calculated for removing acidity, bile, and putrid matter from the stomach. In the beginning of some nervous and putrid fevers, where the pulse is weak, and the stomach loaded with sour, fœtid, yellow or green matter, there is perhaps no medicine equal to it. The heaviness, listlessness, pain in the loins, and head-ache, are generally removed before morning.

Emetic tartar, when it is prescribed with judgment and taken properly, is one of the hest medicines known at this day. I have given it to many thousand patients with the utmost safety, and with the greatest advantage. I prefer it in every case to Jame's fever powder, though a medium composed of the same materials. The operations of emetic tartar may be directed to the stomach, the bowels, or the skin, as the case requires.

Some of the quack doctors mix powdered ginger with the emetic tartar, and call it the ginger vomit. I do not know that this is any injury to the medicine; but some of the low country druggists adulterate it with chalk or magnesia: these articles are only hurtful by preventing the purchasers knowing exactly the quantity they ought to take. It is, therefore, necessary to apply to apothecaries or druggists on whose veracity you can depend.

AN EXCELLENT EYE-WATER.

Take flowers of zinc and white copperas, of each a quarter of an ounce, water half a pint; mix them together. It is used in the same manner as the white copperas eye-water but in most cases is greatly prefferable, particularly in the inflammation of the-lids, and any external or internal excorciation. If it is too sharp, add a little more water to it.

COLD BATHING†

CURES YOUNG CHILDREN OF

Convulsions*
Coughs
Cutaneous inflammations, pimples, and scabs
Gravel

Inflammation of the ears, navel, and mouth
Rickets
Suppression of urine
Vomiting
Want of sleep

IT PREVENTS GROWTH OF HEREDITARY

Apoplexies
Asthmas
Blindness
Consumptions
Deafness
Gout

King's Evil
Melancholy
Palsies
Rheumatism
Stone

---

* And this I apprehend accounts for its frequently curing the bite of a mad dog, especially if it be repeated for twenty-five or thirty days successively.

† Persons of feeble constitution should use it with care, and not remain too long in the bath; after the bath, rub the body until there is a glow over the whole surface.

Yet by this note let no one discontinue the use of baths, for they are approved by all when used properly.

WATER DRINKING
GENERALLY PREVENTS

Apoplexies
Asthmas
Convulsions
Gout
Hysteric fits

Madness
Palsies
Stone
Trembling.

To this children should be used from their cradles.

The best water to drink, especially for those who are much troubled with the wind, is rain water. After it has settled, draw it off clear into another vessel and it will keep sweet for a long time.

ELECTRIFYING
IN A PROPER MANNER, CURES

St. Anthony's Fire
Blindness
Blood extravasated
Bronchocele
Burns or scalds
Coldness in the feet
Contraction of the limbs
Convulsions
Cramp
Deafness
Falling sickness
Feet violently disordered
Felons
Fistula Lachrymalis
Fits

Lock jaws and joints
Leprosy
Menstrual obstruction
Opthalmia
Pain in the stomach
Palsy
Palpitation of the heart
Restores bulk and fulness to
    wasted limbs
Rheumatism
Ring worms
Sciatica
Shingles
Sinews shrunk
Spasm

Flooding	Stiff joints
Ganglions	Sprains, however old
Gout	Surfeit
Head-ache	Swellings of all sorts
Imposthumes	Sore throat
Inflammation	Tooth-ache
Involuntary motion of the eye-lids	Ulcers
	Wens
King's evil	Wasting
Knots in the flesh	Weakness of the legs
Lameness	

Nor have I yet known one single instance wherein it has done harm; so that I cannot but doubt the the veracity of those who have affirmed the contrary. Dr. De Haen positively affirms it can do no hurt in any case; that is, unless the shock be immoderately strong.

The best method is to give fifty, or even a hundred small shocks each time; but let them be so gentle as not terrify the patient in the least.

Drawing sparks removes those tumors on the eye-lids, called barley-corns, by exciting local inflammation, and promoting suppuration.

FASTING SPITTLE

OUTWARDLY APPLIED EVERY MORNING HAS SOMETIMES RELIEVED AND SOMETIMES CURED

Blindness	Corns (mixed with chewed bread and applied every morning
Contracted sinews from a cut	
Cuts (fresh)	Deafness

PRIMITIVE REMEDIES.   141

Eyelids red and inflamed   Sore legs
Scorbutic tetters   Warts

TAKEN INWARDLY IT RELIEVES OR CURES

Asthmas   Leprosy
Cancers   Palsy
Falling sickness   Rheumatism
Gout   Swelled liver
Gravel   Stone
King's evil   Scurvy

The best way is to eat about an ounce of hard bread, or sea-biscuit, every morning, fasting, two or three hours after. This should be done, in stubborn cases, for a month or six weeks.

NERVOUS AND PARALYTIC

DISORDERS ARE FREQUENTLY CURED BY THE COLD BATH, BUT PARTICULARLY

Asthma   Fevers (violent)
Agues of every sort   Gout (running)
Atrophy   Hectic Fevers
Blindness*   Hysteric pains*
Cancer   Incubus
Chin Cough   Inflammations*
Coagulated blood after   Involuntary stool or urine*
　bruises   Lameness
Consumptions   Leprosy (old)
Convulsions   Lethargy
Coughs   Loss of appetite,* smell,*
Complication of distempers   　speech,* taste
Convulsive pains*   Nephritic pains
Deafness   Palpitation of the heart
Dropsy   Stone in the kidneys

Pains in the back, joints, stomach
Rheumatism
Rickets
Rupture
Suffocations
Surfeits (at the beginning)
Sciatica*
Scorbutic pains*
Swelling on the joints
Epilepsy
Torpor of the limbs, even when the use of them is lost
Tetanus
Tympany
Vertigo
St. Vitus' dance
Vigilia
Varicose ulcers
Whites

But in all cases where the nerves are obstructed, such as those marked thus* you should go to bed immediately after and sweat.

'Tis often necessary to use the hot bath a few days before you use the cold.

Wise parents should dip their children in cold water every morning, till they are three-quarters old; and afterwards the hands and feet.

Washing the head every morning in cold water prevents rheums, and cures coughs, old head-aches, and sore eyes.

MANNER OF MAKING PILLS.—Use syrup or the best castile soap to soften the medicines; and to thicken them, use pulverised licorice-root sufficient to give the proper consistency. When made, dust the pills with the licorice powder. —ED.

VINEGAR AND LEMON-JUICE vapor diffused through the rooms of the sick, is both agreeable and wholesome.—ED.

MODERN MEDICINE,

INTRODUCTION

OBJECTS OF MEDICINE.

From laws of health and sickness learn disease ;
Who studies life, her laws more wisely sees.
The greatest duty of the Healing Art
Is first to know *whence all diseases start*.
A skilled physician will be always sure
To read disease, and somehow work a cure,
A doctor, in whatever patient's case,
Has many blind and knotty parts to face ;
As heat of parts, their posture and their form,
Disease, the patient's strength to breast its storm.
He must change solids, check a growing tumor,
And swellings soften into transient humor ;
Or liquify deposits quickly growing,
And turn them all into some liquid flowing.*

By putting this Medical Counselor in the hands of the People, I believe that all will find in it a True Friend in time of need ; and in parts of it, matters of curious interest, as well.

---

* The School of Salernum was a general favorite among the educated of every class, and looked upon, like Solomon's Proverbs, as a Peoples' Book, useful to all who could appreciate the wide and common sense suggestions as to the conduct of our physical life So universally were its merits recognized and endorsed, that according to Mr. Bandy de Bolzue, two hundred and forty editions of this famous poem have been published, and in almost all languages of modern Europe. By permission of the publishers parts of this poem (as rendered by J. Ordronaux, L. L D., M. D.,) will be found throughout this work.

One valuable feature of the work, which will commend it to every mother, is the separate treatment of diseases *peculiar* to *children*, with the milder remedies; while the other treatments given can be used if desired.

The sections: "Laws of Health," "Cause of Fever," "Symptoms of Inflammation," "Acids and Alkalies," and "Scurvy," should be carefully read, as a help to the understanding and treatment of disease.

Fleshy and jovial persons may pass, unread, the section on Dyspepsia and Blues; and boys, that on page 200. While the girls will find on page 299, a subject of interest.

The wife might to advantage cautiously refer the husband to page 304.

To read and apply the counsel given under the head of "Nervous Affections" will bring comfort and peace to many unhomelike homes.

The reader will find the treatment of Asthma and Granulated Sore Eyes, as given by the Author, of uncommon value.

Every unproved remedy has been carefully excluded from these pages: the effort has been made to get the best treatment and remedies from each and all of the various schools of medicine; not by simple compilation, but from bedside provings.

After referring the reader to the preface for a further explanation of the object and character of the work, it is sent on its mission, to win the confidence of the reader, by its good works.

Trusting that the care of an over-ruling Providence may ever be with the means used for the cure of the sick; I remain the friend of the afflicted.

THE AUTHOR.

MODERN MEDICINE.

288. THE NURSE.

The first matter to be taken into consideration is to see that the sick has a nurse possessed of good common sense, and a knowledge of hygenic laws; for a good sensible nurse without a physician is better for the sick, than a good physician, with a careless uninformed nurse. In all acute diseases a good nurse is more than half the battle won.

289. PURITY OF MEDICINES.

Next to this is the purity of medicines. It is best when possible to prepare your own medicines, for medicines are adulterated at a fearful rate; also old deteriorated drugs are sold for fresh, and of course no benefit is had from their use. There are honest competent druggists; but we have also incompetent and unprincipled men all over the country who enter the business for the purpose of having the privilege of selling intoxicating liquors as a beverage to whoever will buy, regardless of the woe, crime and death they produce. Can you trust such an one to deal out to you medicines on which may depend the lives of your dear ones? Pause and think; then be more careful of whom you buy your drugs, than to whom you trust your bank account.

290. KNOWLEDGE OF THE LAWS OF HEALTH.

It is of great importance that we have a knowledge of the laws of health, if we would know the causes of disease and effect a cure. We will thus be enabled to *prevent* sickness or *check* it in its commencement, and thus save much suffering and expense.

LAWS OF HEALTH.

" Be temperate in all things."

Use pure water as a common drink.

Use such food, both as to kind and quantity as *agrees* with your personal wants; that which *your* stomach will digest easily and well: all food does not agree alike with every individual.

Cleanliness in person and surroundings is of vital importance; without this it is vain for any one to hope for long continued exemption from disease.

A due degree and equalization of heat in the system, with neither a great excess or a great lack is necessary to health.

A just proportion of both ALKALIES and ACIDS is one of the greatest conservators of healthy action in the physical man.

Keep the feet dry and the head cool, and the rest of the body EVENLY clothed.

Be at peace with both God and man; this will give a healthy tone to all the fluids of the body.

Pure air, in fact *pure* everything about and in you, gives pure blood and tends largely to a purer life.

Unrestrained or long-continued grief is destructive; while joy, a happy and sunny disposition is a well-spring of life.

As far as practical live in the sunshine in every sense of the word.

OF SLEEP.

Let your head be toward the north or east when sleeping.

> Six hours of sleep suffices for sire and son,
> Seven hours we grant to sloth, and eight to none;
> In less than seven be all thy sleeping done.
> If nine are needed, take not thou a tenth;
> Conform thy sleep to night's appointed length.
> Should health demand, from morning take a third;
> Better that one from bed all day ne'er stirred,
> Than rob his limbs of their accustomed rest;
> And for such sleep the morning hour is best.
> Sleep in due measure profits every one,
> But through excess much ill is often done.
> Nothing is worse than on the back to lie;
> While prone relieves a cough, yet hurts the eye.
> Tis well to change from side to side at night,
> And, nought forbidding, choose at first the right.
> Upon this side begin thy night's repose,
> And on the left let sleep her season close.

OF FOOD.

If people would more generally cook and eat the old fashioned boiled dinners in place of so much sweet meats and rich pastry, it would diminish the number of invalids full twenty per cent. This, of ordinary diet.

> In cabbage we strange contradictions find,
> Its broth will loose, its leaves in contrast bind.
> But broth and leaves, when used together prove
> A laxative, and thus the bowels move.

In the summer season but little fresh beef should be used, winter is the time for its use; smoked ham or breakfast bacon occasionally, with fowls tame or wild, should be used, in preference, notwithstanding the great and popular prejudice to the contrary; but fresh pork very seldom.

The diet in summer should be largely of fruit and vegetables.

But in winter both meat and fruits should be used freely.

> The winter months are cold and damp and drear,
> And tables then should groan 'neath gen'rous cheer.
> Depletions slight, be strong purgations none,
> And roasted dishes to a turn be done.
> With pepper season every kind of meat,
> And once a month repeat love's urgent heat.
> Who keeps this law, physicians all agree,
> Through life in health unceasingly shall be.

The mode of cooking and quantity eaten is of as much importance as the kind of food.

Never gormandize; if you do, you will surely engender disease.

Again, starving either the sick or those in good health, is nearly as bad as the other extreme.

The system requires the *fruit acids* in the winter, just the same as in the olden times, when the Jews were commanded to lay in store wine or grape juice for the winter season. So, now in the "fulness of time," the necessity for the use of wine has passed away; for in CANNED FRUITS we have an article which is far better, and more than fills the place of the Jewish wines. Yet pure grape juice unfermented is very valuable; canned as other fruits.

No chemical analysis can give all of the *qualities* of food. Nature has made such compounds in the vegetable creation as is best suited to the wants of man, and defies the ingenuity of man to learn her secrets. Chemical food is a failure.

So with a good cook (one of the most useful arts) and a good nurse, together with a careful attention to the calls of the appetite both in sickness and in health, the diet may safely be trusted to common sense; *always* giving *small* quantities at first, and increasing in amount as the case will admit.

> At meals begin with meats, with fluids conclude;
> Eschew such drink as toper hands have brewed,
> Which burn our flesh, yet palate all delude.
> Be bread the last of anything you eat,
> And after meals shun fires that give much heat.
> But supper over, then your time beguile
> With rest, or gently strolling for a mile.

BATHING.

> All bathings shun soon after having fed,
> But when digestion's done, no bathings dread.
> Just fresh from nuptial rite or bath, refrain
> From writing, if your sight you would maintain.
> Fresh from the bath to eat is not good,
> Salt water dries the body very much.

> Fresh, tones it ; that of wells chills by its touch.
> In warm baths never make a lengthy stay,
> Lest inner moisture be increased this way.

The mode as well as the cleansing effect of bathing should be considered and understood by all ; it is of the greatest importance to the animal economy that the skin be kept clean ; for such is the relation of the surface to the entire system that the least derangement in its functions produces effects on the constitution from which spring a multitude of diseased conditions.

The perpetual supply and change of the organic elements of the body, are indispensible to its healthy condition. Much of this exhausted or worn-out matter is found immediately in the capillary vessels and have no convenient chance of exit from the body, except through the pores of the skin. When these are obstructed this vast amount of waste matter is retained in the system. Some of it remaining in the capillaries, while much of it is retained in the circulation, and thus in both instances, giving rise to impurity, irritation and *fever*.

Unless this obstructed condition is relieved, the lungs, kidneys, bowels, nasal passages, etc., must suffer by overwork, and the patient becomes an invalid.

But this is not the greatest mischief that arises from obobstruction of the skin exhalents ; the functions of these organs involve the regulation of the temperature of the body. All these changes in the supply and waste of materials, are attended with a corresponding production of *heat* from supply, and *cold* from waste. As there is no evaporation from the surface while the pores are thus closed, there must necessarily be a rapid accumulation of heat in the system, while the retained broken down tissues prove a source of irritation to the nervous system from this prolific source. It must suffice to say that *this is the most fruitful of all the causes of disease.*

All of these mischiefs may occur at any time when the obstruction of the functions of the skin exists, and the extent of the mischief will be equal with the extent of the obstruction.

Not only does mischief ensue from the afore-mentioned causes ; but from the injury entailed on the nervous system. The skin is more extensively supplied with nerves than any other part of the body. The nerves superintend the manifestations of vitality in all parts of the body. As there is a communication through them to all parts ; whatever nervous depression or derangement is produced, will always be attended with a corresponding depression of the vital energies of the system.

From this cause arises a multitude of manifestations of disease ; such as fever, congestions, inflammations of the kidneys, liver, bowels, lungs, etc., the last generally suffers most : but the *weaker* parts first feel the force of the disease.

It is unnecessary to further particularize on the infinite and endless varied forms of diseased action, that may result in the various parts of the system.

Armstrong has well said:

> " The grand discharge, the effusion of the skin,
> Slowly impair'd the languid maladies
> Creep on, and through the sick'ning functions steal ;

> As when the chilling east invades the spring,
> The delicate Norcissus pines away
> In hectic languor ; and slow disease
> Taints all the family of flowers, condemned
> To fade ; should beauty cherish its own bane ?
> Oh ! shame ! nipt with pale quadrille,
> And midnight cares, the bloom of albion dies."

We possess no means for promoting the natural and healthy functions of the skin *equal to bathing*.

MODES OF BATHING.

Sponging with cold or tepid water is one of the most useful of all baths ; this can be used with benefit and safety in almost all fevers and inflammations, and should always be used in cases of this character as often as required to sustain its cooling effect. It is of incalculable benefit and should *never* be neglected.

THE VAPOR BATH.

This bath is of the greatest utility, and often produces astonishing results as a curative agent in rheumatism, and other acute diseases attended with sharp pains or inflammation. A *moist heat* is peculiarly congenial to the body, it equalizes the circulation.

It is always advisable to raise the heat gradually so as not to increase the momentum of the circulation to suddenly. Keep the head and face cool with cold water.

THE SHOWER BATH AND COLD PACK.

Should be used by *feeble* persons only with the council of a competent physician. When used, it should be in the morning, and the body should be well dried with a rough towel, and sufficient friction used to cause the blood to come to the surface; cause a healthy reaction.

OIL BATH.

Often in cases of a hot, red and dry skin, if the whole body is anointed with sweet oil and then properly washed off with salt-water, it changes the violence of the disease into a milder type.

291. OTHER SANATIVE MEASURES.

In malarious districts especially, a fire should always be built in the "living" room every morning and evening, and as far as possible all should avoid the night air; and when going out at night, or early in the morning, a cup of coffee should first be drank, it is a great protection from the malarious poison. One with an empty stomach is more subject to almost any or all diseases. As a preventative, coffee is far superior to whiskey, brandy, or any other alcoholic stimulant.

When possible eat very freely of good ripe sweet grapes.

Let your bed rooms and bedding be *often* aired and sunned. The "spare bed" has filled a multitude of graves, and will continue its work of death, until people learn *never* to allow it to be slept in until all dampness is removed from sheets and bed-clothes.

Plant sun-flowers near your premises.

If these instructions be followed you may bid defiance to ague with all its train of ailments.

292. RECAPITULATION OF THE LAWS OF HEALTH.

> Deep sadness, anger, or unremitting toil,
> All render human life an early spoil
> To death, and thus they hurry on each soul
> Toward the last inevitable goal ;
> While cheerful spirits, magic·like, will raise
> Life's tone, and thus prolong its term of days.
> But such as unto pompous feasts incline
> In youth, invite a premature decline.
> Who tampers with a flux may loose his life,
> The same with cold, much drink and am'rous strife.
> Food, labor, sleep, when moderate each day,
> Do good,—'tis surfeits hurry on decay.
> To rise betimes, at evening to walk late,
> Keep man in health, contented and elate.

FEVERS IN GENERAL.

It will be necessary to speak more at length on the subject of fever than of any other subject properly belonging to a treatise of this character: for he who is in possession of a correct knowledge of the pathology and treatment of fever, possesses the key to the knowledge of all other diseases.

The word "fever" is a generic term, and as such is used to convey the idea of a class of diseases, which stands as the body and centre of all medical enquiry. Fever is the body of the tree; the branches, other diseases; while the roots may represent (in a large degree) the causes or source from which they are fed.

All along the ages, from Hippocrates down to the present time, fever has held the first rank in pathological enquiry. Nearly all systems of medicine have had their foundations laid, and have been built upon their several author's views of fever. Van Snieten says, "If we except those who perish by a violent death, and such as are extinguished by mere old age, and which are indeed few, almost all the rest die either of fever or of diseases accompanied with fever."

Professor J. Eberle says that, "The history of practical medicine, consists of little else than a review of the doctrines which have successively risen and sunk again, concerning the nature and treatment of fever."

The question is asked, "What is fever?" The endeavor will be made to answer this question in as clear and concise a manner as possible; also to give its proper treatment.

WHAT IS FEVER?

1st. Are the remote causes. These all DIMINISH or LOWER THE VITAL POWERS.

It is the retention or accumulation of the natural heat of the body in that degree which shows the phenomenon of disease, by causes which will be explained as far as possible.

Prominent among which are suppressed perspiration, idiosyncrasy and habits of the patient, impurities and other unfavorable conditions of the atmosphere, marsh miasmata, exposure to damp or cold, poor or insufficient food, extreams of heat and cold, injuries, fatigue, grief, and other long-continued, depressing causes; each giving a different shade or modifying influence to the general fever, when it is fully developed by the immediate or proximate cause; while the specific remote cause or contagions, as those of small-pox, measles, scarlatina. etc., give rise to fevers bearing their names.

THERE IS THE IMMEDIATE OR PROXIMATE CAUSE WHICH IS THE SAME IN ALL FEVERS.

☞ The immediate cause of fever is the loss of a healthy balance between the organs of *supply*, or the process of building up of the various tissues of the body from a *fluid* (the blood) into *solids*—which is a *heat*-producing process —and the organs of *waste*, which is a process by which the solids of the various tissues when they have served their purpose in the body, are changed from *solids* to *fluid* waste material; this is a *cooling* process.*

From this it is evident, that when any of the remote causes of fever checks the process of *supply*, we have a sense of *chilliness;* but when the process of waste is stopped, the result is an accumulation of *heat;* and that first one and then the other of these processes is in excess, and the other at the same time proportionately inactive.

---

* This is shown to be true, by the well known principle in chemistry, that heat is always produced when a fluid is condensed or rapidly made to assume a solid form; and on the other hand cold, or a lowering of the temperature, is always produced when a solid is quickly changed by chemical action into a fluid.

These oscillations from cold to heat and from heat to cold continue to take place, creating the remissions and paroxysms of fever, thus showing the kind or character of the diseased action. This continues until, by the unaided conservative powers of Nature, or through the wisely-directed agencies of art, the whole economy settles down to the natural equilibrium, the healthy balance again, or else terminates in death.

NOTE: Of course the nervous forces, etc., have a modifying influence on all diseased action.

This would lead us to expect as the common symptoms of fever, a quick pulse, unnatural heat of the surface, and a cold or chilly sensation at the beginning.

Boerhave collected together from a great number of authors all the symptoms which had been observed in fevers. He then struck from this list all those which do not appear in every form of fever, retaining such only as, by common admission of authors, and his own observations, were found to be present in all cases; and the consequence was that only three symptoms were remaining—namely: a quick and feeble pulse, preternatural heat of the surface of the body, and a sense of cold and chilliness in the commencement.

The reader will see at once that if the *immediate cause* of *all fever* is the *same*, that the *general treatment* of all will be of a *like character*.

While this is the case, the various modifications and symptoms of the different or specific forms of fever require only an occasional supplementary treatment in addition to the general treatment.

Having this *unity of fevers* well fixed in the mind, will

enable the reader to treat *every variety of fever* with greater confidence and success.

294. SYMPTOMS WHICH GENERALLY ATTEND FEVER.

The groups of symptoms are purposely given after the *three universal symptoms* that the reader may not be confused with their number.

When certain groups of these symptoms predominate, it gives name to that particular case of fever.

The symptoms of all fevers are the same, differing only in degrees of intensity, or the mildness and severity of the cases, and the peculiarities and impressibilities of the constitutions in which they are manifested.

There are, generally, several days of indisposition previous to the attack of all fevers. This is called the premonitory state. This state is known by loss of appetite, lassitude and general restlessness, or sometimes an unusual drowsiness; more or less disturbed sleep; with a disinclination to action of any kind.

"In the onset of fevers there are more or less of chilly sensations, attended with flushes of heat. Sometimes these chilly feelings run into agues, or shakes; but with the advance of disease, or as the fever rises, the chilly sensations subside.

In the commencement of all fevers, it rarely happens that there is not more or less of aching pains over the whole body; predominating, however, in particular portions, such as the head, back and limbs. Sometimes this pain is exquisitely distressing; but it is apt to moderate as the disease progresses, or in the language of the schools the reaction becomes established.

In the onset of all fevers, there is more or less of lassitude, weakness, or sense of weariness; in some instances it is considerable, and this weakness continues to increase as the diseased action advances.

In the onset of all fevers there is more or less of increased thirst; sometimes it is intense. This symptom also continues as an attendant on most fevers, but it is apt to moderate as the disease progresses. Although the tongue, lips, and even fauces, or throat, may become more dry in such cases, the desire for drink springs more from the unpleasant sensations of dryness, or suspended secretions in the structure just named, than from the actual cravings of the stomach; since patients in this situation take only a swallow or two at a time, while in some other conditions no amount that could be retained by the stomach would satisfy the desire.

In the onset of all fevers, there is more or less loss of appetite for food; sometimes an aversion for food, amounting to nausea, or sick stomach, without vomiting; at other times with vomiting, and occasionally attended with both puking and purging; and it is well known that there is little or no appetite for food during the progress of the fever.

In the first and earlier stages of all fevers, except in the lowest grade of cold plague, yellow, putrid or typhus fever; and even sometimes in these, the pulse is apt not only to be quicker than natural, but irritated and fretful; giving oftentimes the delusive appearance of increased force, when, in reality, the force is diminished, and the circulation, in such cases, labors to make up in frequency or speed, what it lacks in force. But as the disease advances, and the patient becomes weaker, the pulse becomes still quicker, weaker, and smaller, making very feeble resistance to the pressure of the fingers, and evincing clearly a want of force.

In the first stage of all fevers, the tongue becomes coated with a whitish coat, or fur. This, however is not always discoverable in the onset of the disease, but becomes observable in the course of a few days; and if the fever is not checked, but suffered to run into another stage, then the tongue is apt to assume a yellowish-brown appearance. But should the fever be of a lower grade, or more malignant character—particularly should it partake of the putrid form, such as yellow fever, putrid fever, or the like—then the tongue is apt to assume a dark brown or black color. The casting off these different colored coats from the tongue should ever be considered as a favorable omen. In the last stages of all fevers, the tongue, teeth and lips are apt to become exceedingly dry, and oftentimes incrusted with a dark, stringy, gummy matter, showing a yet greater accumulation of the broken down tissues in the circulation, this is an effort of nature to expel from the system by the mucous surfaces what should have passed off by way of the cuticle (outer covering of the body). In some more rare cases, the tongue assumes a different appearance from any of those that have been mentioned; that is, it assumes a glossy, red appearance. This is apt to take place after many days' continuance of fevers, or about the last stages of the disease, or after relapses, showing the almost exhausted state of the vital forces. Whenever this state of the tongue occurs, we hav generally found the cases to be more tedious and more dangerous.

In some cases of fever there is a roaring sound, or tingling in the ears. This is most common in typhoid, typhus or nervous fevers, but it does, occasionally, occur in any form of fever.

Delirium is a common and a very distressing symptom in all fevers. This particular symptom is more the offspring of peculiar organization than the nature of external causes; some individuals, and some whole families being prone to delirium under fever, from whatsoever causes or under whatever type; while other individuals are capable of passing through any form of fever without much, or any, very appreciable derangement of the functions of the organs of thought.

In the first stages of all fevers there is apt to be a kind or grade of disturbance of mind, not amounting to delirium; that is, a slumbering or dozing condition, that the patient is more cognizant of than his attendants. This symptom, indicating but a slight departure from the normal state, is but of little consequence.

But as the disease progresses in its stages and grades of prostration, the delirium assumes a different and more distressing character; the observing attendants now discover a considerable derangement in the faculties of the brain, of which the patient himself is not conscious. He frequently lies to some extent in a state of stupefaction, often hard of hearing, so much so, that it is difficult to arouse him, and when aroused to make yourself comprehended by him; indifferent to the exterior world, he makes you short reponses, and drops into his slumbers again. In more advanced stages, you find him picking at his fingers and at the bed-clothes, sometimes entirely stripping himself of all covering; again we hear him uttering incoherent words and manifesting a desire to be up and going about, when he is not even able to stand on his feet. In such patients we are apt to witness a creeping or twitching of the muscles of the arms and legs (what is called sub sultus tendinum by the doctors), the voice becomes weak and unnatural, with a wild, vacant expression of the countenance.

There is still another form of delirium, which also occasionally attends the last stages of fever, and indicates a more dangerous condition than the one just mentioned; this is attended with incessant watching, the patient's sleep leaves him, his hearing becomes even more acute than natural, he is startled at every sound, and has imaginary and for the most part distressing images continually passing before his eyes. In all other particulars his condition is pretty much the same as in the state just before mentioned.

Now, as the delirium and muscular agitations or twitchings increase in proportion to the duration of the disease and the physical prostration of the patients, these results must be ascribed to the general exhaustion and consequent debility of the whole physical organization, and not to an increased action of the brain, the heart, the blood-vessels, or any other part of the body, as a large portion of the medical faculty would have you to believe.

In fevers, generally, we observe some more or less sense of heat in the stomach and bowels; we observe, also, an increased color and more scanty secretion of urine.

In the last stages of all fevers we occasionally witness involuntary discharges by stool and by urine; and in all forms of fever these are alike unfavorable symptoms, though not necessarily fatal, as individuals have recovered after such discharges from all manner of fevers.

In *all fevers* the salutary changes, that is, *the crisis*, is usually announced by the occurrence of very *much the same phenomena;* that is, nearly all the cases yield under a return of the salutary action of the depurative functions, particularly that of the skin.

Although the circulation of the blood has been long understood, and the indications to be drawn from the action of the pulse long studied, and many grave, lengthy and learned discourses been delivered from the rostrum of the schools, yet still are the discrepancies of opinion upon this subject as great, even at this day, as at any former period of time. Such, too, must necessarily be the case, for two important reasons—the first is, that the faculty have not yet agreed even in the abstract, as to what constitutes the propriety of depletion; the second is, the difficulty which the mass of practioners find in comprehending the influence of *physical* and *moral causes*, on the phenomena of circulation in their patients. For want of a clear knowledge of all these things, do the most experienced and observant practitioners make blunders.

In the examination of all cases with a view to the indications of cure, we should never fail to take into consideration all the symptoms of the case, as well as all the circumstances of causation. Among the many signs that leads us to a knowledge of the condition of the sick, there are none more uniformly important than those of the tongue and the skin; the strictest attention should therefore be paid to their state and condition.

As a proof that physicians know less about fever than is generally supposed, and less even than they themselves imagine, we hear them continually disputing about the names of fevers, of a country, a place, or a season, such as typhoid, congestive, and bilious fever—just as though they were all distinct diseases, and to be cured by widely varied remedies. The truth is, that the similarity of the symptoms in the early stages of all fevers is so strikingly great, that it is not a matter of astonishment that they should so often dispute about their classification. Many practitioners prescribe more from the appellation which they shall give to the assemblage of symptoms, than from a knowledge of what is taking place in in the economy of their patients; they as often or oftener form their opinions as to the name or nature of a fever from the season of the year in which it occurs, than from the symptoms which attend it. We have known the fever of small-pox mistaken for that of malignant fever—the fever of measles treated for bilious fever. It is not, then, a matter of surprise to us, to find them mistaking one shade of autumnal fever for another, nor would such mistakes be a matter of much importance, provided they understood the nature of the *fever in general*.

We are accustomed to see fevers assume the character of bilious, in warm seasons and in warm countries, while typhoid fevers are more common in the colder seasons and in the colder countries.

We have thought that fever was like unto man, its prototype, never stationary, always either in a state of progresssion or of retrocession, often varying much in its progress, but never fundamentally changing; often running regularly through the different grades of the particular type it may assume, but now and then running from one type to another—from a higher to a lower, or from a lower to a higher; that is, from the simple intermitting form to bilious or even typhoid at one time, while at another time, from typhoid to bilious, and finally ending in chills and fevers. Such things are not of unfrequent occurrence in cases of relapse.

Notwithstanding our conviction as to the truth of the position, that there is a natural chain of connection which combines all fevers together, yet still do we find it convenient for the purpose of making ourselves the easier and better understood by the reader, to adopt the common classification of fevers."

295. GENERAL TREATMENT OF FEVER.

It is always safest to begin the treatment as soon as the symptoms of fever make their appearance.

If the patient feels languid, with a sense of weakness, all business and cares should be laid aside, and he should rest until he is well. On going to bed, he should bathe his feet in hot water and drink freely of hot lemonade, or if desirable, plenty of cold water. If the appetite is poor, he should fast for one or two meals. Should there be sickness at the stomach, a vomit, by drinking two or three pints of tepid or hot water, and tickling the throat with a feather or the fingers will free the stomach of the offending matter.

Should there be constipation of the bowels, they should

be opened by an enema of tepid water or with a mild cathartic.

If there is a hot or dry skin, or he has a headache, a hot or sponge bath should be used, in which a few spoonsful of salt or soda is dissolved. The temperature of the water should be agreeable to his feelings. If agreeable to him, cold water should be applied to the head, otherwise warm or hot.

Fevers in general should be treated with tepid or warm water; for cold water should only be used when the heat of the patient is very intense, and he is of a robust and vigorous constitution.

He should always have fresh air, and sunlight if possible, with cleanliness.

Never "starve" a fever patient; Many are "starved" into their graves.

These plain directions, if used in the beginning of a fever, will never do harm, but always benefit the patient, and quite often stop the progress of the disease. It equalizes the circulation, relieves the congested condition of the small blood vessels in all parts of the system; permits the waste matter, and with it the imprisoned heat, to pass off through the skin, kidneys, lungs, etc.; each part being thus brought into its normal condition. If there are no complications and sufficient vitality remains, the patient is cured.

If the fever should still continue, then use—

1. In the beginning or chill stage, put twenty spoonfuls of water in a glass, then add fifteen drops of Gelseminum. Give a spoonful every half-hour until the chill ceases, and there is a cessation of the symptoms of fever, and perspira-

ion commences. Then a dose every two or three hours. If a few doses should produce no appreciable effect, increase the dose to three or ten drops. If there should be a dimness of vision after giving the larger doses, stop giving the medicine for a few hours. This stops a general or uncomplicated fever in a short time. This effects all that it will ever accomplish the first day ; if the patient is not relieved by that time, then take:

2. Tinc. Aconite one dram, tinc. Bryonia six drops, tinc. Belladonna six drops, put them in an ounce vial, and fill the vial with alcohol.

Dose : One drop for every spoonful of water put in a glass, and give a spoonful from the tumbler every hour until the force of the disease is broken. Effective.

3. Or, the 2d Homeopathic dilution of Acon., Bry., and Bell. Mix. One drop in a spoonful of water every hour.

4. Or, Take a dose for two nights in succession of the pills called the General Catharctic Pills. The formula for making them will be found in the appendix.

5. Or, Fever Root (Sterospora) twenty or thirty grains every half hour until the force of the disease is broken, or until a free perspiration ensues ; then extend the time to three hours.

This generally effects all the good it will do in the course of two or three days in most fevers.

6. Or, Gelsemin one grain, Sulphate of Quinine ten grains, Prussiate of Iron, ten grains; mix and incorporate them well together.

Dose : one grain every hour until relief is had.

7. Or, Tinc. Aconite, two drops, tinc. Gelseminum, forty

drops, water six spoonsful, Sulphate of Quinine eight grains. Shake well when used.

Dose: one spoonful, in sweetened water, every two hours.

8. Or, Fever Drops.

Quaken aspen Bark, Gentian, Capsicum and Prussiate of Iron (all pulverised) of each half a dram; Carbonate of Ammonia 2 drams, Alum 1 dram. Add four ounces of water, and shake the bottle well before each dose is given. The bottle should be tightly corked.

Dose: one teaspoonful every two hours for a grown person, children in proportion.

This is a safe remedy, and of extraordinary value. I have found it reliable as a general treatment in all types of fever. All that is necessary to give confidence in these general remedies is to use them as directed, from one to seven days; the length of time depending on the severity of the case.

9. When in fever or other diseases there is irritation of the mucus membranes, Yellow Root (Hydrastes Canadensis) is a sovereign remedy. A tablespoonful to half a pint of water. Dose: From one drop to a tablespoonful. Water extracts its virtues; it is safe to give it in almost any quantity. Chlorate of Potassa: a teaspoonful to half a pint of the Yellow Root Tea increases its usefulness. Generally three or four drops of the tincture of Nux Vomica to the above mixture will give wonderful satisfaction in the same direction. In Typhoid Fever this is very good.

10. Tincture, or the juice, of Garden Lettuce is good to be given where there is restlessness. Dose of the tincture, from ten to sixty drops; of the juice, what will lie on a silver five or ten cent piece.

11. Sulphate of Atropia.

The Homœpathic preparation (3d) of Atropia is the best remedy for restlessness generally attending fevers. Dose: what will lie on a silver half-dime; to be given every hour until quieted.

12. When Morphine is given it should be combined with Sulphate of Quinine : a quarter of a grain of the first to a grain of the latter.

The doses given are for an adult or grown person; children should take proportionately smaller doses, according to age. (See table of doses.)

13. The patient should be sponged off with a wet towel; and then rubbed dry with a hot, dry one.

This should be repeated as often as necessary. If the surface is cold, use warmth instead ; such as mustard water, hot bricks, vapor bath, etc. Be sure to keep the feet and hands warm, and the head cool.

14. Always remember that there is a condition of debility, that the patient needs tonics both in food and medicine and that animal food is more easily digested than vegetable.

In all cases the patient should have good, plain, wholesome food, anything that is easily digested and agrees with his taste; such as boiled rice, broiled ham, new country cream diluted with sweetened water, hot roasted potatoes, toast, and baked fruits, grapes sweet and ripe, and in some cases oysters, eggs beat into a froth with loaf sugar, etc. The drink should be good pure water, coffee, tea or lemonade made palatable with sugar or milk.

15. Never starve a fever patient by feeding on slops: a well man would soon starve to death on such food as is generally given fever patients.

Better by far that these rules for washing, feeding, giving sunlight and fresh air, etc., without medicine, than medicine without these ; but it is best when both are used.

>   Slender in Spring thy diet be, and spare ;
>   Disease, in Summer, springs from surplus fare.
>   From Autum fruits be careful to abstain,
>   Lest by mischance they should occasion pain.
>   But when rapacious winter has come on,
>   Then freely eat till appetite is gone.

295. CONTINUED FEVER, INFLAMMATORY FEVER, ETC.

Should be treated as given for General Treatment of Fever. In all cases of fever or acute diseases, with fever or inflammation, it is safe to commence the treatment with No. 2, as in the list of remedies. (Page 169.)

296. BILIOUS FEVER.

The predominating symptoms are bilious vomitings and other gastric troubles, with sharp, pungent heat, brownish heavy-coated tongue, high-colored scanty urine, etc., almost from the beginning.

TREATMENT.

Same as general treatment of fever.

And in addition, give a dose of the general cathartic pills, and if need be a second, if the bowels are not freely moved, or Nux Vomica (3d,) Sulphur (3d,) ten grains each, Peruvian Bark (red) 2 grains; to be well mixed in a morter.

DOSE.—One grain (or what will lay on a silver half dime) every two hours.

Or, mercurius vivum 30 grains, Peruvian bark (red) 2 grains, Podophyllin two grains, to be mixed in a morter.

DOSE.—Two grains every two or three hours.

Or, in some extraordinary cases a dose of blue mass, with quinine one grain, at night; followed in the morning by oil or magnesia is very good.

Or, the wild cherry bark bitters as given in the Appendix.

297. TYPHOID OR GASTRIC FEVER,

Gets its name from the peculiar symptoms predominating as described in the section on Symptoms of Fever, viz: Great protsration and weakness of the system early delirium, dark coated tongue, the fever is not nearly so high, deep or profound sleep or sleeplessness, pain in the head or back, the bowels are much affected, and rose-colored spots on the skin, etc.

TREATMENT.

Same as general treatment of fever in Section No. 2, or No. 5, or No. 8. And, in addition, put a handful of slippery elm bark in a cup of cold water, use as a common drink.

A. Or, a tea-spoon half full of glycerine in a table-spoonful of water, to quench the thirst.

B. And the following: Ipecac, Nux (3,) Hydrastus C. (1,) equal parts. Mix; give two grains every two hours.

C. Or, Valerianate of Quinine five grains, pure Brandy two ounces, mix and give a tea-spoonful in sweetened water every two or three hours.

D. Or, make a tea of Wild Cherry bark, Virginia Snake root, and Valer:n root, one and a half drams each, to the pint of water. And give a table-spoonful every three hours until the patient is well.

E. From one to twenty drops Baptisia Tinctora, (wild indigo) or put half an ounce of the root in a pint of water and make a tea of the same, give a table-spoonful every two hours. It is peculiarly beneficial in this fever.

If there is much twitching of the muscles, give ten drops of the following every three hours, viz: 20 grains of camphor dissolved in half an ounce each of sweet spirits of nitre and compound spirits of lavender.

Camphor is one of our best medicines in all low nervous diseases.

Also use

FEVER OINTMENT.

F. Equal parts of tobacco, raisins and lobelia; put into a pan with sufficient lard to cover when melted; simmer for fifteen minutes, strain while hot.

The bowels or any other part where there is inflammation or pain, should be anointed with this; it soon allays the pain and greatly benefits the patient.

If needed, use the

FEVER LINIMENT.

G. Sulphuric Ether and Aqua Ammonia, of each one ounce, Muriate of Ammonia one-eighth ounce, mix and wet the scalp or any other painful part every two hours; covering the part afterward with a wet folded towel, until the pain, abates. These may be used in any fever.

A STIMULANT FOR FEVERS

H. When a stimulant is needed, Oil Cajuputi is used with much advantage in typhoid and typhus fevers, and malignant scarletina. It is preferable to brandy or any alcoholic stimulant.

DOSE.—One to ten drops on sugar or any bland substance. To be used at any time when there is spasmodic action, or, great prostration.

297. TYPHUS FEVER, (OR SHIP FEVER.)

The symptoms are of a lower type than typhoid.

Use No. 8, the Carbonate of Ammo. preparation as directed, together with an occasional dose of Cajuputi, from the beginning.

If needs be, use Nos. 2 or 3 to control the the pulse, and assist the equalization of the circulation. When needed use the ointment, or Liniment as above.

298. INTERMITTENT FEVER, (AGUE.)
OR
AGUE AND FEVER.

Intermittent fever, is marked by successive paroxisms.

COLD STAGE.

A cold sensation creeps along the back, the nails and lips turn blue, and the skin pale; the surface becomes benumbed. The blood recedes from the surface, etc., and in some cases, there is an uncontrolable shaking of the whole body.

HOT STAGE.

Then follows the hot stage with its characteristic fever.

SWEATING STAGE.

The perspiration flows freely, there is an equalized circulation; and for the time being he is convalesent.

These paroxisms occur at regular intervals; as once in twenty-four hours, (quotidian.) or in two days, (tertian) etc. As all are treated practically the same, it is unnecessary to use space in this book to describe them seperately.

This is a typical fever. All of the symptoms will be found under the head of Symptoms which Generally attend Fever.

CAUSE.

Malaria, the posionous effect arising from marshes, stagnant pools, etc.

TREATMENT.

In the hot stage as other fevers. Then Quinine 20 grains, Myrrh 5 grains, Liquorice 15 grains, Dover's powders 10 grains, Capicum 10 grains. Mix and make into twenty pills.

Dose.—One every two hours, until all are taken.

Or, Cinchona 1 dram, Sulphate of Quinine and Piperine each ½ dram, mix, and make into 40 pills.

Dose.—One every two hours.

Or, a table-spoonful of Salt in Elm water twice a day is found effective in some cases.

Or, Sulphuric acid one dram, twelve table-spoonsful of gum mucilage.

Dose.—A table-spoonful every two hours.

Or, Sulph. Quinine 20 grs., pulverised, Capsicum 30 grs., pulverised Carbonate of ammonia 90 grains, mix in a bottle and add 5 table-spoonsful of water.

Dose.—One table-spoonful every hour in a very severe case with a congestive character, otherwise every two hours

Or, put 20 grains of Quinine in ⅛ ounce of lard and anoint the stomach, arm-pits, spine, bowels, etc.

Or, Frankincense 1 ounce, Capsicum, Peruvian bark, and Nutmeg each one dram, salt ½ ounce. Pulverise fine, and mix, put in a cloth sack, and wear it over the stomach. This has cured many.

Or, six grains of Quinine in a two ounce vial, put five drops of Sulphuric acid on it, white sugar a table-spoonful, fill with water.

Dose.—A tea-spoonful every hour, commencing five or six hours before the chill.

Or, a cold bath in the height of the fever, then go to bed and sweat.

If the liver is torpid, give a dose of the anti-bilious pills on going to bed.

In most cases open the bowels with a dose of the general cathartic pills.

Astringents always assist Quinine.

Corydalis (Turkey Corn) and Hydrastis equal parts, is unsurpassed in enlargement of the Spleen after ague. Drink as a tea.

It is necessary to repeat the medicine for a day or two, six days after the chills are stayed, and thus every sixth and seventh days until the 21st day is passed.

For other remedies see appendix.

> Drink not when needless, eat not out of mood ;
> For thirst and hunger *tonic powers* include,
> While surfeits bring of direst ills a brood.
> Note *when* you drink, that you may not fall ill,
> Note *what* you eat ; drink after baths your fill.
> 'Tis heavy, not light, suppers that give pain,
> As common sense and doctors both maintain.
> Unless compelled, you never should combine,
> At one meal divers sorts of food, nor wine ;
> Curb appetite, and thus prolong your health ;
> Temp'rance, the doctors tell us, laughs at death.

299. INFLAMMATION OF THE HEAD.

Pain in the head, symptoms of violent fever, sensitive to light, starts at every noise, and sometimes vomiting and drowsiness, wants to lie down.

TREATMENT.

Treat as fever.

Keep a towel wet with tepid water constantly on the head, changing as often as necessary to keep the head cool. And do not forget the sponge bath. There is great sympathy between the bowels and brain of children, hence any offending substance in the stomach or bowels should be removed. (See first Part, No. 181.)

DISEASES OF CHILDHOOD.

While children may have any of the fevers enumerated; yet it will be convenient to treat of fevers peculiar to childhood days.

Children are more subject to these, from one to twelve years of age. When caused by over-fatigue, improper food, (a common cause) long exposure to the direct heat of the sun, worms, sitting on the damp ground, etc.

SYMPTOMS.

Face red, head hot, throbbing of the veins of the neck, short quick breathing, rapid pulse. With some there are head symptoms; starting at every sound, unable to bear the light, etc.

TREATMENT.

Sponge off the whole body in tepid water, cool water to the head; give Gelseminum or (Aco., Bell., and Bry.,) (2d,) as directed for "General Treatment of Fever."

See that the bowels and stomach are cleansed of offending matter, by enemas or a gentle purge.

If the skin is very dry and hot anoint with sweet oil, bacon fat, goose grease, or oil of almonds.

Be careful that ALL food is well PULVERISED before given the child; and the above will be all the treatment generally needed.

NOTE: Scarlet fever, measles, chicken pox, scarlet rash, etc., are classed among the erruptive, zymoti c, or ferment fevers. The remote causes are a specific eontagion, a specific inflammation of the skin and mucus membrane attends them, and generally having the disease once frees the individual from the danger of a second attack. Like all fevers there is a period intervenes between the reception of the poison and the development of the disease.

301. SCARLET FEVER.

Generally, its first visible effect is on the mucus membrane, often causing vomiting, followed by *violent heat, sore throat*, and rapid pulse, with strawberry tongue and other fever symptoms. Soon the rash appears on the arms, neck, breast and the face; color deepening until the height of the disease is reached, when it gradually fades and becomes of a faint brown or yellowish tint, when the scurf or outer skin scales off in patches, or there is what we call desequamation.

TREATMENT.

Tinc. of Acon., Bell and Bry.; of each six drops put in an ounce vial, fill with alcohol, shake

DOSE.—Put from three to six drops (according to age of child) of this mixture in half a tumbler of water, give a teaspoonful of this every hour at first, and as the disease abates extend the time to two or three hours. A dose of this given once or twice a day will generally protect children from the disease.

It is generally beneficial to anoint the whole body with sweet oil, bacon fat or the "bathing drops," afterwards sponging off with tepid water.

Inhalation from brown sugar boiling is good.

A wet pack about the throat often proves of benefit.

2. In case the eruption is unyielding for a number of days, give

Rhus Tox., Apis and Sulphur (equal parts) ; a dose every hour or two alternated with the first drops.

3. If there is much swelling of the throat,

Gelseminum, Muriatic Acid, Opium, or Mercurius are the proper remedies. To half a pint of Yellow Root (Hydrastis Canadensis) tea, put a tea spoonful of the Chlorate of Potassa, and rinse the throat often.

4. If there should be swelling of the glands about the neck, give

Silicia, Calcarea carbonica, and Baryta carbonica (equal parts), or Lachesis, two or three time a day. Continue the Aconite, Belladonna and Bryonia mixture during the whole course of the fever, extending the time between doses as the pulse may indicate.

5. If at the close of the disease there should be a tendency to dropsical swellings, give Rhus Tox. and Apis mellifica, alternated with Bryonia and Arsenicum (equal parts), every three hours until the child is well.

The anointing and sponging should be continued while the fever lasts.

Give bland but nutritious diet.

They who WILL HAVE a more heroic treatment, may use the following viz :

The vapor bath in the beginning.

Then use the " Stimulating Liniment" over the whole surface of the body. Keep the throat constantly bathed with the liniment.

During the whole course of the disease give a spoonful of the following mixture, every two or three hours, viz.:

Ten drops of the tincture of Gelseminum, Sulphate of Quinine, what will lie on a silver dime. Water ten teaspoonfuls. Mix.

Put a wet towel folded over any painful part and cover it with a dry one.

If there is constipation, use enemas of tepid water, or give a dose of oil.

If there is a suppression of the urine, give from five to ten drops of Sweet Spirits of Niter in a spoonful of Uva Ursi tea every three hours until this is corrected.

When the throat is sore, the "heroic treatment" says:

Take one teaspoonful of red pepper, and the same of Hydrastis, Myrrh, Sumack and pure table salt, pour over them a pint of boiling water: strain. To this add half a pint of good vinegar: sweeten with honey. Of this when cold, from a half to a teaspoonful is to be taken every half hour, and the throat frequently gargled with it.

In this as in all diseases, what has been said on sanitary measures and diet should be borne in mind.

302. MEASLES (Rubeola).

The principal dangers to be apprehended from measles are improper treatment and too early exposures after convalescence; These avoided, and there is but little danger to be feared. It is contagious, epidemic, and seldom attacks the same person more than once. Generally prevails towards spring.

About a week after exposure to the disease, the symptoms of a bad cold are noticed, with a loose, hoarse cough, and fever.

About the third or fourth day, the rough, dark-scarlet-colored rash makes its appearance; first in the throat, face, neck, body, and by the third day it has extended to the lower extremities. It disappears in the same order.

TREATMENT.—Acon., Bell. and Bry. in equal parts. Six drops of the mixture in half a glass of water. Give a spoonful every hour or two, alternated with Puls. If the cough is troublesome give Phos. and Caust. If VERY hoarse, an occasional dose of Spongia and Kali bichr., may be given. A dose of sulphur every other morning, even in convalescence, will be a benefit.

If the eyes are much inflamed, Cal. c., Apis and Euphras. If there is great weakness, China, Rhus tox., Ars. or Nux.

Diet, same as fever.

Sometimes it is a benefit in bringing out the eruption, to give warm Camomile or Saffron tea, with a warm foot bath.

As a preventive, Puls.

303. MUMPS (Parotitis).

The patient should be careful not to expose himself in any way, lest it "fall" on other parts, and produce mischief; otherwise it is not dangerous, but passes off in a few days.

A few doses of Mercurius three times a day is all that is needed.

If there is fever, treat as simple fever. If it should "fall," Puls., Silic, Baryt. c., Lach.

The Magic or Snuff Ointment will materially ease the unpleasant feeling if used on the swelling.

304. CHICKEN POX (Varicella.)

It requires but little medical treatment. Treat as simple fever, or Rhus. tox., Apis, Sulph.

Light diet. It will soon disappear.

305. HIVES—NETTLE RASH, (Urticaria.)

These are irregular white elevations with itching; the wheals are enlarged by scratching, over-heating or cold; often changes from one part to another; if it recedes to the internal organs, there is difficult breathing, *sick oppressive* feeling, this is the only danger attending it; it is not contagious.

TREATMENT.

(Apis and Rhus Tox.,) (Sulphur and Calcarea,) three times a day.

If much fever, Acon., and Bell.

Mercurius, China and Podophyllum two or three times a week. (See Page 266.)

Rubbing the parts with the Magic Ointment, bacon fat or Liq. Chlo. Soda (clear or diluted) gives relief. Fresh pork in summer should not be used.

A fruit or grape diet is best.

> Stripped of all skin, deprived of all their seed,
> Grapes are of highest use in times of need,

> They soothe the swollen liver's angry heat,
> And cool the bile in its own ardent seat.

306. CROUP (Angina Trachealis.)

The mother who has once been aroused in the dead hour of the night, by that *hoarse-barking* or *loud-brazen* cough, can never forget the sound or be deceived as to the character of the disease.

There is difficult rapid breathing, wiry pulse, thirst and other fever symptoms: The child throws his head back in the effort to breathe.

In some cases the child is apparently well during the day, and worse at night; but if left unchecked it will return in more frequent paroxysms, and the wheezing and rattling will become so loud and harsh as to be heard in all parts of the house.

If the child has proper treatment, the breathing will become easier, the harsh rattling softened, and the phlegm loosened.

"But the worst cases are those which come on insidiously, the child being a little indisposed for two, three or four days, *little* or *no cough*, or a *mild muffled cough;* but *very hoarse speech* or *entire loss of voice*, the croupy symptoms come on gradually. Cases of this character are fraught with most danger."

When the false membrane fills the throat, breathing ceases.

Children subject to the croup should avoid exposure at night, in rainy weather, but especially over-heated rooms. Flannel should be worn at all times save in the hotest summer months.

TREATMENT. (Mild Remedies.)

1st. Put the feet and hands in hot water, rub them with your hand, dry with a hot towel, cover the child well; anoint the throat and breast with the Magic Ointment, rubbing and heating it in.

Give (Acon., Bell., and Bry.,) alternating with (Spongia, and Kali Bichro.,); six drops in twelve spoonsful of water.

DOSE.—One spoonful every quarter or half hour, *extending* the time as the child is relieved; when perspiration starts, and the cough is "loose," only give an occasional dose of the *latter*. Now give an occasional dose of Sulph., and Calc. carb., or Phos. If costive use an enema.

DIET.—Fresh warm milk and cold water, if three or four years of age, a little broiled-ham-gravy on *well-cooked* rice.

If on the following night the returning symptoms demand, repeat the treatment.

HEROIC TREATMENT.

2d. Or, a teaspoonful of Croup Syrup every ten minutes in warm water until *free vomiting* follows: anoint the throat and breast with the Magic Ointment, and give a mild cathartic. For Croup Syrup see Appendix.

3d. Or, Alum mixed with sugar equal parts, given in half to teaspoonful doses every ten or fifteen minutes until it vomits; the stomach should be free of milk.

4th. Or, equal parts of Tinc. Lobelia and Blood-root.

DOSE.—For a child two months old six drops, ten months old ten drops, two years a teaspoonful, eight years a dessert-spoonful, to be given in warm tea every fifteen minutes until free vomiting ensues.

Use *hot* water and ointment as in the first. Repeat as often as required by the returning symptoms.

In six hours give from one half to two General Cathartic pills. For recipe for G. C. pills see Appendix.

5th. Or, give equal parts of Lard and Molasses in teaspoonful doses until there is free perspiration or vomiting; anoint the breast, feet, etc., with Goose Oil; the throat with Ammonia Liniment.

While all of these are good; the 2d and 4th can hardly be surpassed. The hot foot and hand bath, etc., should be used in each treatment; and cold water as a drink.

307. WHOOPING COUGH (PERTUSSIS).

This is a contagious disease, and appears as a common cold for a week or two, when it assumes that pecular rapid cough, followed by a deep whooping sound.

TREATMENT.—Should there be much fever, give Acon., Bell., and Bry., as for measles, until the fever is subdued. Then (Drosera and Ipecac.), (Belladonna and Cupr. M.), alternately, every two or three hours.

Put a Belladonna plaster on the spine between the shoulders; from a quarter to half an inch wide.

Vaccination is generally of benefit. Red-clover tea is good for the cough. Prevent exposure to damp or cold winds.

Infants are liable to strangle with the phlegm; they must be constantly watched. A change of air is of benefit.

DIET.—As for fever, also (see part first).

308. WORMS. (Helminthia.)

There are three general species of worms found in the human body. The Round worm, Thread worm, and Tape worm. The symptoms indicating their presence, are multitudinous. The only sure sign, is their presence in the stool or ejection from the mouth. A voracious, then poor appetite ; changing symptoms, fœtid odor of the breath, lower eye lid dark, grinding of the teeth during sleep, swelling of the abdomen, convulsions, picking of the nose, itching of the anus, diarrhea, glaring mucus, etc., give room to suspect the presence of worms.

TREATMENT.

Acon., and Cina if there is fever, convulsions, tossing starting or crying out during sleep, vomiting, etc. Mercurius and Calc., in difficult teething, diarrhea, or constipation.

2d. (Ignatia., Cina, and Bry.,) drop doses in water three times a day. For itching in lower bowels in addition to the above, give an injection of a teaspoonful of Beef's Gall mixed with two ounces of sweet oil, or lime water and milk equal parts.

If there is fever, give fever drops.

3d. Or, Santonine. Dose, from one to three grains rubbed up with sugar, to be given twice a day until the worms are removed, followed on the fourth day by a physic. For further remedies see No. 282.

309. CHAFING OF INFANTS.

Powdered Nut Gall mixed with five times its bulk of

starch, or starch alone dusted on the parts will generally be all that is needed.

Or, ten drops of Hamamelis in a spoonful of water, and bathe the parts.

Frequent bathings will be a good preventive.

310. CRYING, WAKEFULNESS AND COLIC OF INFANTS.

> The nettle to the sick one, slumber brings;
> Checks qualms, and need of all emetic things.
> From painful colics patients may be freed
> By eating honey which contains its seed.

Improper food given infants is often the cause of their crying, restless condition; as well as the eating of green fruits by the mother. The stomach of an infant is very tender, and was never intended by nature to receive any stronger food than the mother's milk. When the stomach is ready to receive solid food, teeth are given for purposes of mastication; and in the same ratio should food of a more solid character be given infants. All manner of bowel complaints, (colic, diarreha etc.,) and even spasams may arise from this cause; mothers cannot be too careful on this point.

Never feed or physic a new born infant if it is *possible* for the mother to give it the breast. It had better go without food for twenty-four hours after birth than to suffer for years from colic, diarrhea, etc.

TREATMENT.

See that there is no external cause for its crying, from the pricking of pins, tight bands, chafing, or hunger. Then if the belly is hard and swelled, the knees drawn up toward the abdomen, and its cries are sudden and sharp, colic is the cause. Give Chamomilla aud Colocynth; alternated with Jalap and Calc. carb., every half hour until quieted.

If there is a fever, give a dose or two of (Acon., Bell. and Bry.)

It there is *wakefulness*, give Cina, or a dose of the first as above, on going to bed.

Or, give Nettle, Chamomilla or Catnip tea.

> 'Mong spices, fennel, as 'tis known full well,
> Hath power supreme all flatus to expel.

See page *266*, for dose.

311. THRUSH, OR SORE MOUTH OF INFANTS.
(Aphthæ.)

Characterised by the mucous membranes of the mouth being covered with patches of a curd-like coat, and often there is an inflamed condition of the mouth which if left unchecked may extend through the whole internal canal, causing dirarrhea, and kindred tronbles.

TREATMENT.

Put what pulverised Borax will lie on a silver half dime in a tablespoonful of honey, and give ten to twenty drops four or five times a day : a weak wash of the Borax used three times a day, will answer in many cases. Generally this is all that is needed.

Or, (Chamomilla, Calc., carb., and Jalap,) three times a day.

Or, Nat. Mur. the same.

Other remedies ; Merc., Ars., Sulph.

For Dose, etc., see page 266.

312. WETTING THE BED, OR INCONTINENCE OF URINE.

It may be congenital, an injury, or from worms. Avoid fluids and fruits after noon.

TREATMENT.

Cannabis S. and Merc. of a morning, and Canth. and Bell. at night. If worms ; Cina is the remedy. Sometimes it requires large doses of Canth. and Bell. in glycerine to effect a cure, ten to twenty drops of the Mother Tincture for each ounce of glycerine, given in teaspoonful doses three times a day. This last for children and adults, *not* infants.

Sleep on a hard bed, rise early, and dress warm.

313. SNUFFLES OF INFANTS.

Anoint the upper part of the nose with sheep's tallow, or magic ointment.

Give an occasional dose of Nux.

314. SORE EYES OF INFANTS.

The sav'ry aniseed the stomach cheers,
And human sight improves as well as clears.
The sweeter kind all others overpeers.

In an acute case of sore eyes, bathe them with the nurse's milk, or Chamomilla tea with a few drops of laudanum in it then use the Hawkeye Salve the size of a pinhead twice a day.

Or, (Apis, Euphrasia and Calc. carb.,) every three hours until relief is had ; at the same time use the *Chamomilla* tea as a wash.

For making Hawkeye Salve see Appendix.

315. SLOW GROWTH OF INFANTS.

Saffron 'tis said, brings comfort to mankind
By giviug rise to cheerfulness of mind.
Restores weak limbs, the liver also mends,
And normal vigor through its substance sends.

(Chamo. and Cina,) (Calc. carb. and Jalap.) mix. These cause the bones to take more nutrition, and the child is soon stronger.

A dose of the first at night, and of the second of a morning.

Dose, see page 266.

316. DIPHTHERIA.

The *Slate colored* patches in the throat mark it as diphtheria. It is a blood poison, there is lasitude and other fever symptoms, breath often offensive, glands of the neck enlarged, offensive discharge from one nostril, and sore throat. The treatment must be prompt and energetic.

TREATMENT.

Put pulverised Chlorate of Potassa and Hydrastis (Yellow Root), a teaspoonful of each in a tumbler of water; gargle and swallow a teaspoonful of this every half hour, and anoint the throat with the Magic ointment, heat it well in, then put a bag of hot salt around the throat, keep the salt as hot as it can be borne. It is of advantage to give a spoonful of gruel, elm water, beef tea or other bland nourishment before each dose of medicine.

During the day give three or four doses of the Carbonate of Ammonia mixture as given for fever. If the fever is very high give a few doses of Acon., and Bell.

This treatment never disappoints me.

Or, put a teaspoonful of sulphur in a glass of water, stir well before each dose, for it is hard to mix; with this, gargle the throat often; and use the ointment as above.

This is reputed to be of great value in this disease.

Dr. G. E. Shipman says: " Though attended with inflammation, it should be borne in mind, that it is a disease in which there is *great prostration* of the *vital* powers, hence means must be used to keep up the strength of the patient. As regards the treatment, as soon as any appearance of cold presents itself give *Camph.*; if the symptoms of cold are very marked, put the patient to bed, and give a teaspoonful of the solution every fifteen or twenty minutes, till perspiration ensues; after that less frequently. If the throat should show signs of inflammation, give *Merc. dulc* (2d,) every two hours, and gargle or swab out the throat with *Camphor* solution every hour between. Put a thin slice of pork in a thin bandage and tie it about the throat.

This treatment will usually suffice for the milder cases. For those which are more severe, give *Merc. iod.* (1,) with the *Chmphor* gargle as before, while the fever is high; and *Ars. iod.*, when the fever begins to subside. Where the disease seems to be a combination of Croup and Diphtheria, *Kali bichrom.* is the remedy, a dose every hour or two.

A small quantity—as much as will lie on a five cent piece—may be put into some boiling water, and the child may inhale the fumes, it may be kept boiling near the child, and thus keep the atmosphere impregnated with it, the fumes of *Camphor* are also useful sometimes in these cases. When the false membrane is thrown out, the danger is not passed; the strength must be kept up by nourishing food; unhappily at such times, the stomach often rejects what it should retain; here the *Ars. iod.* is of use, but the other remedies sometimes come into play—these must be consulted in the Repetory. Other children should be kept apart from the patient as much as possible, and a piece of *Camphor* kept in the mouth, by those who must be exposed to the breath of the patient, may prevent infection."

Or, Phytol., Lachesis, and Mer. iod. Mix. A dose every hour or two.

317. TEETHING.

The first teeth that are cut, are the lower front ones, then the two upper, in from four to eight months from birth.

This important process is often accompanied with diarrhea or "summer complaint," feverishness, convulsions, fretfulness, and wakefulness.

If flannel is worn, and the mother's milk only, given, and the child often bathed, any serious difficulty may be avoided.

Treat diarrhea, fever, or any of the diseases enumerated above as they are treated under their several heads.

If the gums are very badly swollen, and the tooth is nearly through, lancing the gums will give relief.

If the bowels are constipated, the juice of prunes may be used.

Calc. c., and Silica are useful in slow dentition.

> Prunes cool the body and bowels move.
> To all, in many ways, a blessing prove.

318. CONVULSIONS OR SPASMS.

The symptoms are so well known, that it will be unnecessary to give them here.

Cause.—Nervous irritation; by anger or other excessive passions of the nurse, worms, *uncooked fruit*, meat, candies, inflamed gums, fright, hereditary predisposition, and diseases of the brain.

Loosen the clothes, especially about the neck, give plenty of fresh air, put the feet in *very* warm water, apply cloths wrung out of cold water to the head. Be calm, do not get excited, and as soon as the fit is off—

Give (Ignat., Bell. and Sulph.,) alternately with (Acon. and Bry.) Give injections of warm water freely, and repeat until there is a *free evacuation* of the bowels.

Also see Part First of this book, No. 41.

During the Summer months the following has been almost always successful when used, viz :

Bromide of Potassium from two to ten grains in an ounce vial filled with water; give of this from ten drops to a teaspoonful every half hour at first, then extend to an hour, and two hours, as relief is had. At the same time give of the following, one powder every two hours.

Calomel one grain, Camphor two grains, Calc. c., ten grains; rub them well together, divide into twelve powders. Let them be worked off with Oil or some other mild physic. If there is no inflammation of the bowels an occasional drop of Oil Cajuput on sugar, will prevent a recurrence of spasms.

Use injections, and bathe the feet as directed.

319. DIARRHEA OF CHILDREN.

> These are foremost—bitter, acid, salt—
> Acids cool, bind : and styptics have this fault ;
> The oily, sweet, insipid though they be,
> From all extremes will keep the body free.

Infants in health generally have two or three discharges from the bowels in twenty-four hours; while it is fraught with danger to *entirely* check the discharges in diarrhea, an excessive number of operations is to be guarded against, by *removing the cause.*

CAUSE.

Indigestible food, over-heating or violent passions of the nurse, unripe or decaying fruits, unmasticated food of any kind, exposure to cold or damp by nurse or child, limbs or bowels *thinly* clothed, (this is a fruitful source of this disorder,) impure air, teething, and other like causes.

TREATMENT.

Avoid the above causes. As much as possible infants should be confined in their diet to the mother's milk. children to well cooked rice, breakfast bacon or broiled ham well masticated, milk porridge, farina, etc., the *juice* of prunes or sweet ripe grapes may sometimes be allowed.

Sponge off and *rub dry.*

Give (Ipec., Calc. c., and China) equal parts after each operation.

Or, (Chamomilla and Puls.) alternated with (Ipec., Podo. and Calc. c.,) the same.

Or, (Merc. Cor., Colch., and Coloc.,) the same.

Scorch corn or corn starch and cook it in new milk or chicken broth, use as a drink.

If caused by worms use the treatment given under that head. Use injections of starch with a few drops of tinc. of Hamamelis in it, or slippery Elm mucilage in new milk, in which put a few drops of Paragoric.

If more vigorous means are required, then use the treatment given for Cholera Infantum.

320. CHOLERA INFANTUM.

Let air you breathe be sunny, clear and light,
Free from disease or cesspool's fetid blight.

There is generally vomiting and purging, the discharges are generally thin, light colored, green or bilious. There is more or less fever and emaciation. It may run from a few days to three months, in the latter case the little sufferer becomes almost a skeleton.

Carefully read what is said under the head of "Diarrhea of Children," and observe the counsel given. You cannot be to careful of the child's diet.

TREATMENT.

(Verat. a., Arsen., Cupr.,) alternate with (China, Ipec., and Calc. c.,) every one two or three hours, owing to the urgency of the case; give an occasional dose of tinc. of Camphor, from *two to ten* drops on sugar.

The following more heroic treatment is quite effective in its results.

Ipecac, Camphor and Calomel one grain each, Morphine one quarter of a grain, Sugar of Milk twelve grains. Mix and divide into twelve powders; give one powder every half hour or hour until four have been given or relief had; then give the following,

Leptandrin six grains, Quinine three grains, Camphor one and a half grains, Ipecac three fourths of a grain. Mix, and divide into twelve powders, one to be given every two or three hours for several days, then once or twice a day for a week, and the cure will be permanent.

Or, the Carbonate of Ammonia mixture No. 8, P. 170.

Or, oilCajeput and tinc. Camphor equal parts.

Dose.—One drop on sugar as often as needed.

Or, Rhubarb and Peppermint one ounce each, Capsicum and Cinnamon bark one dram each; cover with hot water, steep and strain, add Carbonate of Potash three drams, white sugar a quarter of a pound, Brandy equal to the liquid in amount.

DOSE.—For a child from thirty drops to two teaspoonsful every two or three hours.

Use injections of Chammomile, Slippery Elm and Smart Weed tea; or as in diarrhea.

See also Diarrhea and Dysentery of adults.

321. FLUX AND DYSENTERY.

Constant straining, evacuation scanty, mucus mixed with more or less blood, whitish egg like discharges, like the scrapings of entrails, or green mucus, all more or less streaked with blood.

The colicky cutting pains are very severe.

TREATMENT.

Keep quiet and in a recumbent position as much as possible. Drink but little, but when necessary use the *scorched* corn or elm water; eat thickened milk boiled, rice well cooked, the juice of sweet grapes, and this with care; adults may *carefully try* crackers and cheese.

Use injections as for diarrhea; also a warm hip bath often does good.

1. If there is much fever give (Acon., Bry., and Bell.,) for a few hours, then give (Coloc., Colch., and Mer. cor.,) every hour until better, then extend the time between doses. If there is a *burning* pain, give a few doses of Ars.

If a Heroic treatment is desired, then use

2. The Dysentery Pills as given in the Appendix, they are a speçific in this disease.

Or, 3. Give (according to age) from ten drops to a teaspoonful of Compound Syrup of Rhubarb or nutralizing mixture every two hours, until the bowels are well-cleansed. Then give extract of Bayberry, Camphor, and Carbonate of Ammonia equal parts ; to be pulverised fine and mixed· Dose. from one to five grains every time the bowels move.

Or, 4. The Diarrhea and Dyssentery Mixture as given in the Appendix.

The treatment of all diseases not found under the head of "Diseases of Childhood" is the same as for adults, in proportion to the age. See table of doses.

To prevent mothers from being unnecessarily frightened by a type of sickness peculiar to boyhood, the following is inserted :

"The various diseases which attack boyhood are worthy of study. When a man is sick he just takes to his bed and says no more about it. If he is very sick, he is calm, placid, and tractable. He takes his medicine without a murmur, and sips his chicken broth with becoming gratitude. If he is only a little ill, he is harder to take care of than half a menagerie. He can do more grumbling and find more fault without cause than the most ambitious hypochondriac could hope to put into a lifetime. But a boy is generally taken with alarming, if not fatal, symptoms at about half-past eight a. m., when school opens at nine. The fond mother looks at his tongue, feels his head to see if it is bursting with fever, and then tells her husband to drop in at the Doctor's on his way down, and ask him to hurry lest he be to late. At exactly nine o'clock the boy's pulse grows steadier, and there is a calmer look in his eyes. The extreme lassitude which betokened typhoid has passed off, and by the time the boys in the class are wishing that they had been smart enough to have the same fever, he is sitting by the river's brink and putting a worm on the hook. Boyhood is peculiar."

322. YELLOW FEVER.

> Whene'er this dreggy humor doth invade,
> The pulse is hard, the renal stream delayed,
> The skin throughout assumes a dusky shade,
> Gloom reigns, and fear and dreams of darkest hue,
> And acid belchings, taste and spit, ensue.
> The lightest sound borne on the atmosphere,
> Whistlings and ringings causes in the ear.

This fever is common to warm climates. The remote causes, are a specific poison, cesspools, and *decaying filth* under a hot sun. Immediate cause, the same as other fevers.

SYMPTOMS.

In mild cases very much the same as Bilious Remittent Fever.

It commences suddenly, with giddiness, headache, chills, pain in the limbs and back, in a few hours there is fever, with throbbing pulse, flushed face, and red eyes, great thirst; tongue coated white, pain and oppression in the stomach; as the disease progresses and by drink these symptoms are aggravated; the bilious matter thrown up is acrid. The countenance has an expression of distress and hopelessness. In thirty-six hours there is a gradual decrease of these symptoms for the same length of time, the patient recovers, or after a short period of rest, the attack is more severe than at the first. In the second stage there are *burning* pains in the stomach, straining to vomit, a watery greenish mucilage flakey mass is thrown up. Now the eyes, neck and breast are yellow, and a dirty yellow tinge extends over the whole body. Again in about thirty-six hours there is an *intermission* if death has not closed the scene. While many die, and some recover during the progress of this stage of the disease, others pass into the third stage; all of the symptoms of this stage show the *vitality almost exhausted ;* rapidly sinking pulse, *brown* or *black coated tongue,* vomiting severe and constant, the ejected matter is brown or black, called *coffee ground* or *black vomit,* mixed with a glary liquid ; intolerable burning in the stomach, discharges from the bowels green or black, and now uncontrollable bleeding often occurs. Then follows cold extremities, hiccough, convulsions, and sometimes delirium and death.

It sometimes proves fatal in a few hours. It generally runs its course in from five to seven days. The symptoms most unfavorable are black vomit, convulsions, and delirium.

TREATMENT.

(Ipecac and Aconite) the mother tincture ; drop doses, alternated with (Bell. and Bry.,) every half hour: give ice cold Chamomile tea made VERY SOUR with LEMON juice, also Ice Cream freely ; as soon as the stomach will bear it, give No. 8, (P. 170,) in place of the medicines first directed.

Bathe the surface of the whole body with the following : put a spoonful each of Mustard, Soda and Capsicum, in a basin, rub them together, pour a pint of boiling water over them, add as much whiskey.

If there is much pain in any part, use the Fever Liniment. (P. 175.)

Read Section 14, P. 171. Use enemas when needed to keep the bowels open. Pepper, Salt, and *burned* corn well pulverized may be used freely in the food.

When the patient is convalescent use the General Tonic. See Appendix.

The Author was a subject of the Yellow Fever pest at one time, and speaks from experience.

323. SMALL-POX—(Variola.)

Small-Pox is a very contagious disease : The confluent is a more severe type than the distinct.

SYMPTOMS.

A *bruised pain* in the *loins* and *back*, a white furred tongue, vomiting, *dots* which give to the touch the sensation of *shot under* the *skin*, together with *general fever* symptoms. After exposure to commencement of the attack, is from one to three weeks : the eruption appears on the third day, first on the forehead. The fever subsides when the eruption is completed. The boil-like pustules have a *button hole* appearance in the *centre* and fill with matter of a very disagreeable odor, and in from seven to nine days they break, and form scales which fall off in a few days.

If not properly treated there may be secondary fever when the pustules are ripening, if so, there is great danger of *after diseases* which may be of a serious character.

The Author having had an extensive experience in the

treatment of Small-Pox, can truthfully say that it is less to be feared than Measles, when properly treated.

TREATMENT.

The patient should have plenty of fresh air without much light, and be kept moderately cool. When disinfectants are needed, use Copperas dissolved in water, hot roasted Coffee, or Carbolic Acid in water, and Chlorate of lime, or sliced Onions changed often will be of benefit.

Or, Salicylic Acid is a disinfectant of uncommon value.

Take one ounce of Cream of Tartar, dissolve in a pint of hot water, when cold drink an occasional draft until it is all used. This is almost a specific, it seems to destroy the poison throughout the whole system. If there is much eruption paint the *face* and *hands* every day with the Carbolic Acid Ointment ; (see Appendix for Recipe) and on the patient's recovery, the ointment and scabs will come off like a mould from the face and hands, leaving no scars. Fat *Bacon* may be used for anointing when the other can not be had.

If other treatment is desired the following mild remedies will be best, viz :

(Acon., Bell., and Bry.,) for the fever, alternated with (Rhus tox., Sulph. and Apis) for the eruption, a dose every two hours.

If there are Typhoid symptoms, treat as typhoid fever.

The diet should be nourishing, but no pork or bacon should be allowed. (See fever.)

To disinfect the clothing, and bedding, bury them in the ground in some sheltered place ; the room and furniture should be washed in a weak solution of Carbolic Acid, after having been fumigated with burning Sulphur.

As a preventative *vaccinate* with PURE *vaccine*. Some claim Black Cohosh as a complete preventative.

324. VARIOLOIDES (Modified Small-Pox.)

Should be treated the same as Small-Pox.

325. INFLAMMATION.

When the conditions for producing inflammation exist, either general or local, the *general* character and treatment of disease thus produced is in a *general* way the same; modified in some measure by the cause and the tissues or parts inflamed.

It may be a restorative or sanative action, as in surgery, where *healing* by first intention is well known to be the result of inflammation.

When violent or excessive, or the *vitality* of the patient or part is low, it is destructive in its action. Some tissues are more liable to take on imflammation than others. In inflammation of the *serous membranes* there is effusion of *serum* and *lymph;* of the *mucous* membranes there is effusion of *mucous;* it produces softening of the brain, abscess in the liver, and gangrene in the lungs.

The *arteries* are not easily inflamed by injuries; the *veins* are; yet the *general* characteristics of inflammations are the same in all parts of the system.

SYMPTOMS.

There is *redness, swelling, increased heat and pain*, and if extensive, it is generally ushered in with a chill.

In inflammation of the *fleshy* or *fibrous* structures, there is a *dull, gnawing, aching pain.*

In the *mucous membrane*, (as the *throat, stomach, bowels bladder, eyes,* etc.,) the pain is *stinging* or *burning.*

In the *serous* membranes, (such are the *plura* or internal lining of the chest, the membranous *covering* of the *brain,* the *exterior* covering of the *bladder* and *kidneys,*) the *pain* is very *sharp*, cutting, and almost unendurable.

When the *nervous* structure is inflamed, the *pain* is *keen, jumping, quick,* and extremely severe; such is the *toothache,* and neuralgia.

These symptoms will give a close observer a pretty correct knowledge of the *parts* involved in any given case of inflammation.

GENERAL TREATMENT.

Equalize the circulation by cold, hot, or vapor baths, and give (Acon., Bell. aud Bry.,) or the Sweating Powders or Pills; (see Appendix) drink of cold water or hot teas, as suits the patient best, until free perspiration starts; avoid any kind of exposure; keep the feet as hot as possible; hot fomentations, or a cold compress to the painful parts; or a thorough application of the Magic Ointment, a poultice of Stramonium leaves, or the Cajeput liniment will relieve the pain. The cause if known, must be removed or avoided.

If complicated with other diseases, the treatment given for these must be used as well.

326. ERYSIPELAS (St. Anthony's Fire.)

Erysipelas is easily distinguished from other forms of disease: it may be *recognized* by the *redness* and *swelling*, *burning-heat*, with *intolerable itching*. The *eruption differs* from that of measles by its *rapid spreading*, and the *abruptness* of its *edges;* from scarlet fever by its *marked inflammatory character*. It is often ushered in with characteristic symptoms of fever.

When the cause is constitutional it is called idopathic, and generally affects the *face* and *neck;* when it follows injuries, traumatic. Any cause that will produce a depraved or corrupt condition of the system, be it malaria, exhaustion from exposure to excessive heat or cold, intemperance, wounds, badly ventilated rooms, or any other cause which depresses the vital forces, may be followed by erysipelas.

TREATMENT.

If there is much fever give (Acon., Bell. and Bry.,) a dose every hour until the fever abates; then give (Rhus tox, Apis, and Sulph.,) every two or three hours until the patient is convalescent. Use a *mild* purge, or an enema of tepid water.

Paint the parts with tincture of Iodine.

Or, give a mild cathartic, or an enema at once; then give (Acon., Bell., and Bry.,) a dose every hour until the patient is better, then extend the time between doses. Apply Fever Liniment to the head as directed on page 175. Bathe the feet in hot mustard water. Sponge of the whole body as in fever. Sometimes the Acitate of Ammonia in teaspoonful doses given every hour or two until perspiration ensues, is of great benefit.

Or, use a lotion of Verat. vir.

Or, apply a liniment made of equal parts of Lime-water and Linseed oil.

Or, Cranberry poultice.

Or, the application of fresh blood or flesh to the parts often effects a cure.

If there is great exhaustion and danger of gangrene, use the following:

Soda bicarb., and Carbonate of Ammonia, each five grains, Salicylic acid twenty grains, water two ounces.

Dose.—From a teaspoonful to a tablespoonful every two hours.

Unless there are complications requiring the attention of a Surgeon, the above treatment is all that will be need.

Diet.—The same as for fever. (See inflammation.)

AFFECTIONS OF THE HEAD.

For the convenience of the non-professional reader, the various diseases which follow this heading, have been arranged in *groups* according with the parts affected by the *most marked* symptoms; commencing with the head, then the face, throat, chest, stomach, bowels, etc., and thus until all *parts* are included.

328. INFLAMMATION OF THE BRAIN. (Phrenitis.)

In this disease there is violent headache, some fever, redness of the face and eyes, the patient is very sensitive to light and noise, full pulse, looks wild, grits his teeth, talks at random, more or less vomiting, the pain is most severe when the membranes covering the brain are the parts affected.

TREATMENT.

The same as brain fever.

329. HEADACHE. (Cephololgia).

If wine give headache, water drink alone,
To follow tippling fever's very prone.
Should crown of forehead heated be and ache
Light frictions of these parts let patients make,
And with infusions hot of Morel lave ;
'Tis said from headache they have power to save.

Headaches seldom exists independently of some other derangement or disease. They may be located in a single part, or the whole head may be involved. They are usually present in all fevers, and sometimes they are periodic, returning every seven, fourteen or twenty-first days ; a foul condition of the stomach, and constipation of the bowels, over eating, drinking spirits, exposure to intense solar heat, colds, excessive study, nervous excitement or prostrations, or derangement of the monthly periods generally produce headaches.

It is called *sick headache* when there is nausea or vomiting.

TREATMENT.

First learn the cause in each case, and remove or avoid it as far as possible.

Give of Atropia (3rd) what will lie on a silver half dime, alternated every half hour or hour with drop doses of the mother tincture of Aconite : if not relieved by three doses of each, then wait an hour and repeat. Or on the first indication of sick headache give from two to ten drops of the mother *tincture* of *Ipecac* or Lobelia in a gill of water, repeat the dose every half hour until relieved.

Or, give (Nux v., Iris v., and Apis) a dose every half hour or hour until relieved, then extend the time between doses ; in sick headache double the dose. If there is much fever alternate (Acon., Bell., and Bry.,) with the last.

If caused by the monthly periods, (Apis and Puls.) alternated with (Sepi and Calc. c) every two or three hours.

If from dyspepsia, give (China, Sulph. and Calca. c.,) at night, and (Nux v., Iris v., and Apis) each morning. For dose see P. 266.

Or, tinc. of Valerian, tinc. of Blood-Root and Yellow-Root equal parts ; dose from ten drops to a teaspoonsful in water, or chared corn (powdered), one or two tablespoonsful. Use the Liniment as in fever (P. 175) or the Stimulating liniment in the same manner. Bathe the feet in hot mustard water just before "going to bed" at night.

Or, eating salt until thirsty, then drinking *hot Tea* or cold water until it causes a sweat is very effective in many cases when caused by cold. If coffee or tea causes headache in any given case, stop their use.

> Coffee to *some* gives sleep, to some unrest,
> Headache relieves, and stomache when oppressed ;
> Will monthly flow and urine too procure-
> Take it slow-roasted, each grain picked and pure.

230. APOPLEXY.

Apoplexy may be distinguished from epilepsy by the *ab-*

sence of *convulsions*, and the presence of loud snorting breathing.

Congestion of the brain or excessive loss of blood either one may be the exciting cause.

It is often followed by a paralyzed condition of one side of the body.

Persons of intemperate habits, corpulent, short necked, sedentary habits, or aged persons, are most liable to suffer from this disease.

TREATMENT.

1. Remove the clothing from around the neck, give plenty of fresh air, elevate the head, use cold or ice water freely on the head and neck, rub the whole body with Anti Spasmodic Liniment. (See Appendix). Pound the palms of the hands and soles of the feet with the flat of the attendants hand.

2. Aqua Ammonia and Camphor should be inhaled.

Then take Sulphuric Ether one ounce, Camphor gum one dram, Oil Cajeput one dram, mix and shake well : of this give from five to twenty drops in syrup or water ; to be repeated every fifteen or twenty minutes until he is revived.

Or, 3. Carbonate of Ammonia twenty grains, one qnarter of it in half a gill of water; repeat in fifteen or twenty minutes if not revived. If he is unable to swallow, use the same by injection in elm mucilage.

4. Or, use the vapor bath.

MILDER TREATMENT.

Or, after conforming with directions in section 1, of this subject, give (Acon., and Bell.,) alternated every ten or fifteen minutes with (Ignatia and Sulph.)

When sufficiently recovered the bowels should be moved

by an injection or a dose of the General Cathartic pills. Avoid ALL EXCESSES.

331. SUN STROKE. (Coup de Soleil.)

Should be treated the same as apoplexy.

332. EPILEPSY OR FALLING SICKNESS.

> Headache, catarrh the violet dispels,
> And *falling fits* and drunkenness expells.

There is a loss of consciousness during the fit, convulsions and foaming at the mouth, followed by sleep.

TREATMENT.

The same as apoplexy or spasms.

To prevent return of the fits, electricity properly used is often of benefit. See "Electricity."

Or, from three to ten grain doses of Bromide of potassium, dissolved in Valerian, blue Cohosh or Skull-cap tea should be used morning, noon and night until there is no return of the disease.

333. SCALD HEAD. (Tinea Capitis.)

The eruption on the scalp may be either a dry scurf, or inflamed pustules which discharge a thick yellowish matter, forming a crust in patches over the scalp.

TREATMENT.

Cut the hair short, and anoint the diseased scalp three

times a day with the Cutis ointment, or Carbolic Acid ointment. For making which see Appendix.

Give (Rhus tox. and Apis) in the morning, and (Sulph. and Calc. c.) at night.

334. INFLAMMATION OF THE EYES. (Ophthalmia.)

Ophthalmia may be confined to the lids and mucous membrane of the eye, or the inflammation may implicate the whole globe of the eye. There are several varieties with names to correspond with the exciting causes. But the *general* character of the inflammation is the same.

SYMPTOMS.

The eyes are *red* and *swollen*, with a stinging or burning pain ; there is a feeling as if there were grains of sand under the lids, and in some cases the heat and pain extends into the head, producing great suffering ; there is a flow of scalding tears and an intollerance of light ; *granulations* form on the mucous surface ; also there is generally a secretion of mucus gluing the lids together.

CAUSES.

Much bathing, venus, blust'ring winds and wine,
And wounds, or any serious blows in fine.
With lentils, pepper, mustard, also beans,
Garlic and onions—by such hurtful means,
With too much labor amid dust and smoke,
Weeping, or watching fires, we thus invoke,
With long exposure to noon-day sun,
The direst wrongs that can to sight be done.
But vigils are, by far, more noxious still
Than any form of single-mentioned ill.

Contagion, exposure to sudden changes of temperature, injuries, weeping, syphilis, over-taxing the eyes by much reading at night, scrofulous taint, a debiliated condition of the general system and *malaria*.

TREATMENT.

First remove any known cause of irritation.

2d. If there is *much* inflammation give a dose of the General Cathartic Pills every other night until three or four doses have been used. Bathe the eyes freely four or five times a day with the following lotion: To one pint of Chamomile tea put a tablespoonful of Laudanum and as much pure Sulphate of Zinc as will lie on a silver three cent piece, let it be warm or cold as is most agreeable to the patient. Or,

Elder flowers or bark steeped in soft water, and a few drops of Laudanum added and used the same as the above, will soon reduce the inflammation.

As soon as the active inflammation is subdued, apply the HAWK EYE SALVE three times a day, and the *granulations and soreness will soon disappear.* (For making the salve see Appendix.) If caused by malaria or other diseases which effect the general system, use the treatment for that particular disease in addition to the above.

Or, make a strong decoction of Witch Hazel bark, Golden Seal, and Lobelia; evaporate to the consistency of syrup; apply with a camel's hair pencil, twice a day, and use the General Tonic. (See Appendix).

Or, give (Apis, Euphra. and Calc. c.); three or four times a day, and bathe the eyes with Cal ndula and Hamamelis lotion.

> Fennel, vervain, rose, celandine and rue,
> Cure filmy eyes and give them light anew.
> From each a potent eye-wash may be made,
> To strengthen them when light begins to fade.

The diet should be generous, but not stimulating; bathing, fresh air, moderate exercise and everything calculated to improve the general health should be practiced. **Also** see P. 192, and Part first.

335. FALLING OF THE EYE-LID. (Ptosis.)

This complaint should be treated by a competent surgeon. Nux. v. (3rd,) and Atropia (3rd,) will be found of benefit; give the first in the morning, and the latter at night. Electricity, or bathing in cold water is useful.

For the mode of application see " Electricity."

336. STY ON THE EYE-LIDS. (Hordeolum.)

TREATMENT.

Carbonate af Ammonia in from one to five grain doses three times a day.

Or, (Puls., and Apis,) in the morning, and (Euphra., and Calc. c.,) at night. If much inflamed apply an Elm poultice. White bread poultices should NEVER be applied to the eyes.

337. FOR A BRUISED OR "BLACK EYE."

Apply raw beef or Arnica lotion.

337. WEAK AND FAILING SIGHT.

> Swallows, to their blind young, with celandine,
> Restore, 'tis said their wonted vision fine ;
> And Pliny writes that if this be employed,
> Vision returns to eyes of old, destroyed.

TREATMENT.

If it is from the humors of the eye loosing their convexity by reason of approaching old age or other causes (Far-Sightedness or Presbyopia), the eyes should be gently pressed from the outer angles when washing the face ; or use rubber cups on them once or twice a day, but never to the extent of causing pain. If from too great convexity of the cornea, (Short-Sightedness, or Myopia), then close the eyes and use soft pressure on the front part of the eyes by the same means If from other causes, use the Hawk Eye Salve and the general tonic as given in the Appendix.

338. CROSS-EYES. (Strabismus.)

CAUSES.

Short or weak-sightedness, spraining or injury of the muscles of the eye, by purposely looking cross-eyed, looking sideways at a light while being rocked in the cradle, or watching rapidly passing objects from a car in motion.

A child of the Author was injured in this way, from the last mentioned cause ; but by carefully abstaining from over-taxing her eyes, now in her budding womanhood it is scarcely perceptible, and is no detraction to her person, but rather adds piquancy and interest. If any interference is required, a competent surgeon should treat the case ; but as a general rule a *mother's care* is better than the *surgeon's knife*, for by a surgical operation there is danger of the eyes being turned in an opposite direction. Carefully comply with the laws of health as given under that heading.

Other diseases of the eye might be added, but it would make the book unnecessarily voluminous.

339. INFLAMMATION OF THE EARS. (Otitis.)

The pain accompanying inflammation of the ears is a combination of nervous and mucous membrane pains, as described on page 206.

The patient generally complains of "Ear Ache :" Often there is puriform mucous or pus discharged from the ear.

CAUSES.

The same as other inflammations; or scarlet fever, measels, or scrofula may produce it.

TREATMENT.

Tobacco or Stramonium smoke blown into the ear, or hot fomentations to the ear gives relief. A few drops of Camphor, Lobelia, the oil from Peach meats, or a Pepper corn put in a thin cloth and dipped in sweet oil and put in the ear relieves the pain.

If the disease is attended with an excited pulse, give (Aco. and Bell.,) every hour until the active inflammation is subdued ; then [Puls., and Calc. c.,] twice a day.

If there is a discharge from the ears, give [Hep. Sulph. Silic. and Puls.,] morning and night.

340. DEAFNESS. (Surditas.)

CAUSES AND TREATMENT.

To sleep soon after having taken food,
And excesses when frequently renewed,
With drunkenness—all these in turn appear
To dull, betimes, the sharpness of the ear.

When caused by the bursting of the drum of the ear, an artificial drum is the only remedy.

When caused by inflammation; for treatment see inflammation of the ears.

When from cold; give Pulsatilla.

For catarrhal deafness when the throat and glands of the neck are implicated; give (Silic., Puls. and Hep. Sulph.,) twice a day.

If caused by measles or small-pox give the same. When the auditory nerve, and bones of the ear are not materially injured, the Audiphone is worthy of a trial.

If by the accumulation of ear-wax, syringe the ears with warm suds; the finest of soap should be used.

Rattlesnake, Dog, or Skunk oils, mixed with equal parts of Sulphuric Ether, and a few drops put in the ear three or four times a week, has cured many.

Stew Sweet clover in Goose oil, then strain and drop three drops daily in the ear: this is often of much benefit.

341. CATARRH. (Coryza.) CHRONIC. (Ozœna Narium.)

Catarrh is treated under this heading for the reason that it has become so popularly known as a disease of the nose, though not strictly such; the mucous membrane of the throat, lungs, or even the whole system may be involved as well.

As a rule a sneeze is the first warning nature gives that some part of the body is exposed to a colder temperature than other parts, that the sneezer is catching cold. This is an effort on the part of the system to equalize the circulation, to start perspiration, which if heeded would put an end to the catarrh before it began.

SYMPTOMS WHEN ACUTE.

The same as a "common cold;" there is a sense of chilliness, redness of the eyes, heaviness and pain in the head, the nostrils are obstructed and dry, or, there is a discharge of a thin colorless fluid from them; sore throat, heaviness, slight fever, quick pulse, tongue coated white, appetite poor, and general lassitude.

TREATMENT.

The same as given for "General Treatment of Fever," from page 167 to 169 inclusive.

> Fast well and watch. Eat hot your daily fare.
> Work some, and breathe a warm and humid air;
> Of drink be spare; your breath at times suspend,
> These things observe if you your cold would end.
> A cold whose ill effects extend as far
> As in the chest, is known as a catarrh.
> Bronchitis, if into the throat it flows—
> Coryza, if it reach alone the nose.
> Fresh from the bath get warm. Rest after food,
> Or walk, as seems most suited to your mood.
> But in what'eer engaged, or sport, or feat,
> *Cool not too soon* the body when in heat.

Use the vapor bath at night, and the cold sponge bath on rising in the morning ; and remember that friction or rubbing after a bath is what gives it force or value.

Give Alterative No. 1, of the Appendix ; and at the same time take of Iodine and Carbolic Acid one dram each, Aqua Ammonia and tinc. Camphor two drams each. Mix ; put cotton in a vial and drop therein ten drops of the mixture ; inhale the fumes through the nose, three or four times a day.

Or, take equal parts of Cubebs, Golden Seal, and Hamamelis, pulverise and smoke, forcing the smoke through the nose : and at the same time use the Magic Ointment on the bridge of the nose, between the eyes, and adjacent parts.

Or, use the preparations as a wash for the internal naries which are given for inflammation of the eyes.

Once a day give a dose of the tonic No. 1., of the Appendix.

Give a dose of the General Cathartic pills (see Appendix) twice a week.

Eat plenty of "*roast beef.*"

Or, of the milder remedies ; give, (Puls., Nitric acid and Aurum muriat). Mix. A dose morning noon and night. At the same time put of Merc. cor. (4th) what will lie on a dime, into a glass of soft water, and snuff a tablespoonful of this up the nostrils three or four times a day.

Other remedies : Ars., Merc., Nux., Kali bich., Sulph., and Acon.

342. BLEEDING AT THE NOSE. (Epistaxis.)

To stop, clasp the hands above the head.

Or put a roll of paper under the upper lip; by this means there is a compression of the arteries that feed the bleeding parts.

Or, use Hamamelis, Alum, Tannin, Bearberry, or a Puffball to the parts.

Bathe the face and head with cold water, snuffing it up the nose.

Give a few doses of (Acon. and Hamamelis) if there is fever. See P. 266 for dose.

> Should ever bleeding from the nose begin,
> 'Twil yield at once to the leek smeared within.

For Thrush or Sore Mouth, Disorders of teething, Sore Throat, Diphtheria and Mumps, see "Diseases of Childhood." For adults increase the size of the doses: see table of doses.

343. TOOTHACHE.

TREATMENT.

> Thus treat your teeth when'eer they chance to ache;
> The seeds of leeks, select wisely, take,
> Burn them with sweet frankincense mixed, nor yet
> To introduce some henbane leaves forget;
> Then through a funnel broad allow, forsooth,
> The smoke to be slow drawn into the tooth.

If caused by decayed teeth; moisten a piece of cotton with Oil of Cajuput and Oil of Cloves to which has been added one quarter of a grain of Morphine, and fill the cavity.

Or, Alum and Salt equal parts, pressed firmly into the tooth.

Or, use the Liniment No. 2. See Appendix.

Or, hot fomentations.

When badly decayed they should be extracted.

If caused by cold; give from one to ten drops of tinc. of Gelseminum every hour, keeping the feet in hot water until perspiration starts, lie down and cover warm; after a rest, sponge off the whole body with moderately cool water, and give a dose of the Cathartic pills.

MILDER REMEDIES.

If caused by neuralgia, give (Mezer. Bell. and Platina m.,) alternated with (Acon., and Bry.) a dose every hour until relieved.

If it is Rheumatic toothache; give Tart. emet., Merc. Bell., Arn., or Rhus tox.

If from Dyspepsia, give (Nux., China and Iris v.,) every hour.

If from the monthly periods; Puls., Macrotine, Caulo. Calc. c., or Cham.

Brush the teeth after each meal; after supper pure Soap, Borax, Alum, or Salt should be added to the water thus used.

Electricity and cool baths, used with good sense, will be found of great utility in preventing a return of toothache.

See " Elictricity."

344. OFFENSIVE BREATH.

A fœtid breath indicates an unhealthy condition of the secretions, disease of the lungs, throat, stomach, bowels, or decayed teeth. Indeed so many are the causes producing it that it would be unwise to attempt giving more than general directions for its treatment. When the cause is known, use the treatment given under *that* heading; otherwise, use the following.

TREATMENT.

Salicylic acid and Golden Seal twenty grains each, Carbonate of Ammonia five grains, water four ounces. Mix, and keep the vial well corked. With this rinse the mouth and throat morning and evening, swallowing a spoonful each time.

Remove all decayed teeth, or have them cleansed and filled with gold.

345. SCURVY. (Scorbutus.)

Scurvy is *pre-eminently* a disease of the *general system*, a disease of *innutrition*, and it is only placed in this connection for the reason that the symptoms which first attract the notice of the patient are spongy and bleeding gums.

Although not recognized as such by medical writers, scurvy is a *condition* of the *system* which if properly understood will simplify our knowledge of a large class of diseases, and like fever it becomes the *centre* from which eminates a multitude of ail ments. And when the cause and treatment are understood, what seemed dark and enveloped in mystery before, now becomes obvious. In fact to *all who would successfully treat disease* a *knowledge* of the SCORBUTIC DIATHESIS is of PRIME IMPORTANCE.

SYMPTOMS.

The scurvy is gradual in making its development, there is

heaviness, weariness, dejection, loss of strength, loss of energy, emaciation, pale and *bloated* countenance, *lived* spots on the *skin, spongy gums,* offensive breath, swelling of the legs, bleedings, foul ulcers, offensive urine and stools, old wounds break out afresh, severe *wandering pains* prevent much sleep, the skin is dry, the pulse is small, quick and often irregular, and the tongue is dark coated, but there is no loss of intellect or consciousness.

(A pale color of the tongue with a white fur indicates an *excess* of *acids.*)

These symptoms may exist in every grade, from a type so mild as to scarcely attract the attention, to a state wherein the patient prays for death to end his misery.

CAUSE.

☞ Lack of Nutrition, caused by a DEFICIENCY of ACIDS, for acids through the medium of the lymphatics are used by the "Vital Force" as the means for the decomposition and removal of the waste material which has served its purpose in the animal economy; to be replaced through the medium of the lacteals by fresh nutritive material which for its ultimate reception into the animal organism is dependent on the action of the ALKALINE element.

Hence ACIDS and *Alkalies* are essential elements in the food of *all* animals; and in their proportions there must be a reasonable balance maintained; for on the maintainance of the *balance between nutrition* and *waste* depends the healthy action of the entire system.

We also find the scurvy *diathesis* in Ship Fever, Small-Pox and other diseases where there is a visciated condition of the fluids of the body.

For the supply of Acids, nature has made broad provisions in the form of fruits and vegetables by way of the stomach, and oxygen of the air through the lungs.

The Alkaline element is found in salt and most of the food that we eat, as well as the water we drink.

TREATMENT.

The treatment is obvious; first, Acids should be given to cause the removal of the waste or broken down tissues, then the Alkalies to assist in building up the new or healthy ones: In fact a very large percent of all medicines owe their *medicinal* virtues directly or indirectly to either the Acid or Alkaline element.

Give Lemons, Oranges, Grapes and other fruits; Potatoes and other vegetables; fresh lean meat and milk.

Or, make a beer of the Centaury Plant, Sage, Tansy or Celery and drink of it freely.

To be followed by No. 8, P. 170.

Or, Scurvy Grass tea.

Or, use water made sour with Tartaric or Citric acids as a beverage. Then give the following: of Soda bicarb and Ammonia carb. each five grains, Salicylic acid twenty grains, water two ounces; keep it well corked: Dose, from teaspoonful to a tablespoonful twice a day.

Bathing with vinegar in which pepper has been steeped should be practiced every day. If the weather is pleasant, moderate outdoor exercise should be practiced.

Keep the bowels regulated.

The milder remedies are, Carb. v., Ars., Merc., Sulph. and Rhus tox.

Why should he die whose garden groweth sage.
No other plant with death such strife can wage.
Sage soothes the nerves, and stills a trembling hand,
And sharpest fevers fly at its command.
The beaver, sage, and lavender will bring,
With tansy, and cress, first gifts of Spring,
In trembling limbs, a new awakening.
For guardian Sage is nature' soothing king,

346. QUINSY. (Cynanche Tonsillaris.)

In Quinsy there is inflammation of the tonsils and mucous membrane of the throat, with rapid swelling and throbbing pain of the parts, hoarseness and difficult swallowing; together with general fever symptoms.

TREATMENT.

Give (Acon., Bell. and Bry.,) to control the fever. Then apply the Stimulating Liniment No. 2, to the throat: otherwise it should be treated the same as Diphtheria.

347. GOITRE. (Bronchocele.)

This affection is characterised by the enlargement of the thyroid gland, swelling on both sides of the larynx and trachea, its growth is slow, and generally its only inconvenience is its size.

CAUSE.

An *excess* of *Alkalidity* from drinking water as a beverage which has passed through Magnesian Limestone.

TREATMENT.

Use acids as in Scurvy, freely and continually.

Also, take Iodide of Potassium one dram, water one pint, mix and after having used the acids for two weeks, give a teaspoonful of this twice a day for the same length of time, and thus alternate the two ; first the acids for two weeks, and then the Iodide of Potassium for the same length of time.

Apply externally to the swelling with friction the following mixture :

Equal parts of the soft extract of Yellow Willow, the Stimulating Liniment, and pulverised Salt ; this should be used three times a day for two weeks ; then Iodine Ointment for the same length of time ; thus alternated until a cure is effected.

Of the milder remedies, Spongia, Calcarea, Mer. iod., are the most important.

348. CARBUNCLE. (Anthrax.)

"A CARBUNCLE is a hard and circumscribed inflammatory *Tubercle*, like a boil, usually occurring on the *neck*, face, or back, and is much disposed to terminate in mortification. These tumors are exceedingly painful, and discharge very fœtid, bloody matter. In the centre of the tumor is a black core, which has been likened, by the ancients, to a *burning coal* consuming the surrounding parts, and hence its name."

TREATMENT.

The following seldom fails to arrest its further development when used in the beginning, and if a *thorough appli-cation* of the treatment is given it effects a cure in very unpromising cases.

Take Cream of Tartar half an ounce and dissolve it in a pint of water; drink all of it in the course of the first day; then on each succeeding day until cured, take from one to ten grains of Carbonate of Ammonia every four hours; it should be dissolved in elm mucilage, water, syrup, tea, or made into pills. If left open to the air it will loose its strength, for it is very volatile.

Also use the following as a local application.

Glycerine one ounce, Carbolic acid from fifteen drops to half a dram, mix and force a sufficient amount into the Carbuncle so as to cause it to come in contact with all of the diseased parts; then cover the parts with the lining membrane of an egg, or oiled silk.

Or, use the Cutis Ointment.

Or, the Corbolic Ointment. See Appendix.

Or, Turpentine spread on as a plaster is highly recommended by some authors.

If there is much inflammation use an elm or flaxseed poultice.

Of the milder remedies use (Rhus tox., Apis and Sulph.,) or Arsenicum a dose three times a day, and apply a solution of Carbonate of Ammonia to the Carbuncle, and cover it with oiled silk.

DIET: Give plenty of roast beef, Oranges, Grapes, Lemons, Prunes, etc., but abstain from pork and milk.

<pre>
          FIG-POULTICE will our bodies rid of tumors,
              Scrofula, boils and even pecant humors;
</pre>

> 'Twill surely draw—add poppy-heads alone—
> The splintered fragments from a broken bone.

The following is so completely in harmony with the Author's views, that its insertion here requires no apology.

MEDICINAL USES OF CARBOLIC ACID.

The diseases in which carbolic acid is especially useful are,

1. All that class of local festering, pustulating diseases of the skin which at times are at once so common and so difficult to cure. They include all kinds of pustules, boils, and carbuncle ; sycosis, pustular acne, and festering ringworm.

2. Such strumous sores (especially of the neck) as comes under the care of the physician.

3. Excoriations of the os and canal of the cervix uteri.

4. Phthisis in its second and third stages, and cases of chronic bronchitis accompanied with more or less purulent expectoration. In order to be effective the carbolic acid must be brought into contact with the part to be acted on, and in many cases where it has been found ineffective, the failure has been due to a neglect to insure this contact. In the pustulating and suppurating diseases of the skin it is never sufficient to apply the solution of the acid, of whatever strength, upon or to the outside of the skin. It must always be introduced into the interior of the sore or pustule itself, and so as to come sufficiently in contact with every part of the diseased surface. All cases of boils and carbuncles in their earlier stages can be absolutely aborted and cured, whilst even in latter stages their further increase can be almost surely prevented. For this purpose a very strong glycerine solution should be employed, and it is best con-

veyed into the interior of the pustule, boil, or suppurating spot, by a new quill pen dipped into the solution, and introduced by a rotatary motion through its apex, where a sufficient aperture will generally be found. In carbuncles, which are necessarily larger, and have often several openings, several such inductions may be necessary, or, at a later period, threads of lint soaked in the fluid, may be passed with a probe well into all the sievelike openings. Occasionally, as when the mass is large and solid, a watery solution of the acid may be injected with a hypodermic syringe into various parts of the hardened growth. The same plan of treatment is often quite effective in cases of sycosis, pustular acne, and festering ringworm. (Dr. P. Eade, Lancet June 29, p. 927.)

349. MINOR AFFECTIONS OF THE SKIN.
(Cutaneous Diseases.)

As the most important among the inflammatory and febrile eruptive diseases have been already noticed, and as these minor cutaneous affections all require similar treatment, and are of a general character; it is thought best to give them a brief notice in this place. Notwithstanding their character as minor diseases, *some* of them have always been considered difficult of cure; but to the treatment here given, they generally yield.

350. TETTER.

(Crusta lactea, Humid Tetter, Running scall.)
Eczema appears under different varieties, which are all

non-contagious vesicular scaly affections, and like scald-head appearing on various parts of the body but mostly on the hairy scalp ; around the beard ; under the arms ; on the pubis ; breast ; arms ; hands, and between the fingers, etc. It does not have the honey-combed appearance of Porrigo. It is distinguished from Itch by the vesicles being flatter and grouped together, while in the Itch they are more isolated.

The itching in Eczema is of a smarting painful character, while in Scabies it affords pleasure to scratch the parts.

When Tetter appears in the form of a vesicular eruption in groups on an inflamed base, perfectly circumscribed, and seperated from each other by intervals of sound skin, it is called Herpes ; to this class belong Ring-worm ; Shingles (which is generally about the waist) ; eruptions of the lips or margin of the mouth, and other parts.

In Crusted Tetter ; Running Tetter ; (Impetigo) the pustules are generally grouped in clusters, forming thick yellowish or semi-transparent incrustations ; the skin feels "tight," and the heat and itching is almost *intolerable*. It is useless to occupy further space in an attempt to individualise the various phases of diseases of the skin ; it is sufficient to say that *all crusty, scabby,* or vesicular *eruptions,* with *scaly, rough,* or *chapped skin* attended with *itching* and burning ; on which the crusts soon form again after falling off, covering again the inflamed parts ; all of these and many more, be they called Tetter, Salt rheum, Herpes, Acnea, or by any other name ; may practically have the same treatment, with the best and most satisfactory results.

TREATMENT AND DIET.

1. These should be the same as for CARBUNCLE.

2. Or, Calendula (marrigold), Hamamelis and Vaseline made into a Cerate: this should be used if the skin is broken; if the skin is unbroken the Tinc. of Tamus, or Iodine Ointment may be used to advantage on the diseased parts.

3. Or, equal parts of Sweet Gum and Vaseline melted together and applied to the affected parts.

4. Or, Shoemaker's wax as a plaster.

5. Or, Tinctures of Lobelia, Sanguinaria, Poke root and Princes's Feather equal parts; mix and apply to the parts.

6. Or, equal parts of Carbolic acid and Stramonium ointments, mixed and applied.

7. Or, the Hawk Eye Salve.

8. Or, Ointment No. 3, of the Appendix, applied to the parts, is good.

9. Or, if there is a Syphilitic or Scrofulous taint: take equal parts of Sumach, Hamamelis and Turkey Corn, make a strong decoction and drink of it freely, at the same time keep the sores dampened with the same: Five grains of Iodide of Potassium added to each dose of the above, often increases its good effects.

10. Old ulcers, suppurating and irratative sores on the limbs and various parts of the body, Boils, Fistulous ulcers, etc., should receive the same treatment, especially that part given for Carbuncle.

11. The medicines directed to be used internally, should not be neglected, especially the milder ones.

12. With Sulphur, Orpin mix—bear this in mind—
And add some lime, with yellow soap combined;

> By these in mass commingled well and milled,
> Fistula's cured, if four times it is filled.

DIET and other accessory measures.

The diet should be generous; but stimulating food and drink, or that not easy of digestion should be avoided. Celery, Asparagus, Oranges, Grapes, or other vegetables and fruits of a similar kind, together with *roast* beef and brown bread, is the most appropriate diet. Strict attention should be given to the Hygienic measures recommended in Sec. 290, P. 148.

In diseases of the skin, "Cuticura Soap," is useful; to be used in the bath.

351. BARBER'S ITCH.
(Porigo, Scald-chin.)

The treatment should be the same as that given for Scald-Head, or for Carbuncle.

Or, take cigar ashes dampened with spittle, and thoroughly apply it to the parts diseased. This should be used in this way three times a day.

352. ITCH. (Scabies.)

Itch is a vesicular eruption, caused by the itch mite; the eruption is found chiefly on or about the joints, more especially between the fingers.

TREATMENT.

Give a dose of Sulphur and Cream of Tartar twice or

three times a week ; and apply externally every night Sulphur ointment, (of Sulphur, what will lie on a quarter of a dollar, mixed with an ounce of Cosmoline ;) and wash the entire person each morning ; change the under clothing and sheets every other day ; and the disease will soon disappear.

Or, use the Carbolic Acid, or Cutis Ointment.

353. CHILBLAINS. (Perniones.) OR FROSTED HANDS, FEET, ETC.

This inflammation is attended with a sense of burning, itching, tingling, swelling, and occasionally there is ulceration.

CAUSES.

Constitutional tendency, and exposure or excessive cold, "Frost bite."

FROZEN PARTS should be kept away from the fire, and rubbed with *snow* or cold water, until they are thawed out. If this course is persued there will *be* no Chilblains.

TREATMENT.

The parts affected with Chilblains should be bathed morning and night with one of the followiug lotions, viz : Tamus and Arnica.

Or, Tinctures of Calendula aud Hamamelis equal parts; and if in addition the unbroken surface is painted with Tinc. of Iodine once a day, it will facilitate the cure.

Or, Cosmoline one ounce, Copperas half a dram ; pulverise the Copperas, and thoroughly mix ; then add Armenian bole and Venice turpentine one dram each, oil of Burgamot ten drops ; let them be thoroughly mixed in a mortar.

Apply this ointment to the frosted parts three or four times a day.

BLISTERED FEET and Chapped Hands may be treated in the same way.

Or, rub the feet, on going to bed, with Tinc. of Camphor mixed with Tallow dropped from a lighted candle.

Or, use the Cutis ointment : see Appendix.

(Apis., Rhus tox., and Sulph.,) may be given once or twice a day.

Other internal remedies are Puls., Tamus com., Calend., Bell., Ars., Phos., and Arnica.

354. CORNS. (Clavi Pedis.) WARTS. (Verrucœ.)

First remove all pressure from the corns or warts : Bathe the parts affected at night with water in which has been dissolved a spoonful of Soda for each quart of water used ; and in the morning, bathe the warts or corns with Vinegar.

A good way to remove the pressure is to cover the corn or wart with a sponge, leather, or plaster, with a hole made through the center.

On retiring at night apply cotton saturated with Acetic acid, strong Vinegar or Lemon juice.

Or, a solution of Carbonate of Ammonia.

Or, Willow bark in Vinegar of any sort, when macerated long, dissolves a wart.

Or, take equal parts of Willow bark, Blood root, Carbonate of Ammonia, and Cayenne, all finely powdered, shoemaker's wax enough to make a plaster ; mix, and apply as directed above.

Or, paint them with Tinc. of Iodine twice a day.

Or, bind cotton batting on them, and wet it twice a day with equal parts of Spirits of Camphor and Turpentine.

Or, slake Potash and make into a paste with pulverised Gum Arabic, and apply for five or ten minutes, then remove and apply Sweet Oil.

Or, of the milder medicines, give Thuja (3), or Rhus tox (3), morning and night; at the same time paint the corns or warts with a strong Tincture of either Thuja or Rhus tox.

355. FELON. (Paronychia.) WHITLOW. (Paneris.) RUNROUND. (Tourniole.)

Inflammatory abscesses of the fingers receive various names dependent on the tissue affected. When they are situated in the membrane covering the bone, they are called Felon ; when not so deep seated, but affecting the cellular tissue, they are called Whitlow; when they are seated in the skin near the nail, Runround. These abscesses are in their nature essentially the same as when affecting tissues in any other part of the body. The firm unyielding character of the tissues involved, is the cause of the terrible throbbing pain and slow development.

TREATMENT.

If used at an early stage of the disease, the following will be effective : Put a Fly blister about the size of the finger nail, over the painful part and let it remain there over night.

Or, the following seldom fails even in the advanced stage: let the finger be immersed in weak lye, and gradually heat it; the finger should be held in as long as it can be borne, then take Yellow Clay and make a mortar with a saturated tincture of Camphor, spread this on a cloth and tie closely around the finger; keep it wet with the Camphor for several days; and there will be no grounds for complaint.

The finger held in the smoke of burning sugar, or the steam from a decoction of Tansy, Hops and Wormwood will relieve from the intolerable pain.

Or, Lance to the bone.

356. AFFECTIONS OF THE ORGANS OF BREATHING AND OTHER DISEASES OF THE CHEST.

INFLUENZA. (Catarrhus Epidemicus.)

This is an epidemic arising from a peculiar condition of the atmosphere, and has all the symptoms of a common cold; but the chilliness, heat, sneezing, acrid discharge from the nostrils, headache, cough, pain in the back and limbs, suffusion of the eyes, nausea and depression of strength is more severe and persistent, there is more or less fever, quick pulse, and furred tongue.

TREATMENT.

It should be the same as the treatment given at the onset of fever, under the head of "General Treatment of Fever." P. 167. Sec. 295.

Or, take Galangal root, Carbonate of Ammonia, and Hydrastes a teaspoonful each, put them in half a pint of hot water, take it hot, a tablespoonful every half hour until perspiration ensues; also dilute and use the same as a snuff.

If the cough is troublesome take a spoonful of the Lung Balsam (No. 1,) every hour. (See Appendix.)

Or, of the milder remedies, the same as for Catarrh.

357. A COMMON COLD.

For treatment, etc., see Sec. 295. (P. 167) and Sec. 342. (P. 220.)

Or, eat three or four Oranges or Lemons within the hour, and put on extra clothing.

Or, heat the feet very hot for half an hour, and at the same time eat freely of salt or smoked ham until quite thirsty, then drink all the cold water you want.

Or, brisk exciting exercise until perspiration starts may be taken, then put on an extra garment and cool off gradually away from a current of air.

If there is a COUGH,

Use Lung Balsam No. 1, of the Appendix.

Or, use Galangal root and Yellow Star as a snuff, at the same time drinking hot teas.

Use the vapor and sponge baths.

358. CLERGYMAN'S SORE THROAT, OR LARYN- GITES AND BRONCHITES.

(Inflammation of the Throat and Bronchial Tubes.)

Public speakers and singers are liable to sore throat and

hoarseness; arising from and produced by irritation and inflammation of the Larynix and Bronchial Tubes.

"The symptoms vary very materially; but in every case there is cough, differing, however, in its character and intensity, but always more marked, perhaps, than in other diseases of the respiratory organs. The expectoration, too, differs greatly. At times, it is clear and transparent; at others, very frothy: or, on the other hand, viscid, adhesive, and containing small, white grains, which adhere to the vessel. These particles have been mistaken for portions of pulmonary tubercle, and, therefore, have been supposed to be indicative of phthisis; but if there be any doubt as to their nature and origin, the doubt may be dissolved by placing some of them on a piece of paper, and exposing them to heat. If they are merely sebaceous matter from the mucous follicles of the fauces and pharynx, they will leave on the paper a *greasy stain*, which will not be the case, provided they are tubercular matter from the lungs.

In other cases, the sputta consists of a greenish yellow puriform mucus, which may either form a homogeneous mass, or the matter of each expectoration may remain distinct. Commonly they are devoid of smell, but at times they are insupportably fœtid.

The respiration may not be much affected; but, commonly, it is more or less oppressed, and, at times, seems to be complicated, as it were, with asthma—the difficulty of breathing recurring in paroxysms. Commonly, there is not much pain attendant upon it, unless the paroxysms of coughing are frequent and severe. Nor are the general symptoms usually marked. At times there is no accompanying fever;

but, in other cases, the febrile movement is considerable; the inflammatory affection passing to the subacute form. When this is the case, the nutrition of the system is affected; emaciation takes place, with evident febrile exacerbations toward evening, and all the signs of hectic, under which the individual is gradually worn away."

TREATMENT.

1. Chew and swallow a small piece of the *fresh* root of Stillingia three or four times a day; and at the same time apply the following mixture to the throat, (and if at the same time a drop or two is taken, so much the better): oils of Stillingia, Lobelia, Cajeput and Cinnamon one dram each, Alcohol two ounces, mix; this is almost a specific.

2. Or, use the Cough mixture No. 1, of the Appendix when the cough is troublesome, and apply the above mixture externally.

3. Or, inhale the fumes of the Carbolic acid mixture as in Catarrh; in some cases the treatment given for Catarrh may be used to advantage.

4. Or, Capsicnm, Salt, Gum Myrrh, Hydrastes and Hamamelis equal parts, steeped in equal parts of Vinegar, Water and Honey. Use it at will.

The wet compress and the morning bath should not be neglected.

5. Or, of the milder remedies.

If there is fever, give (Acon., Bell. and Bry.), then (Bry., Phos., and Caust.,) three or four times a day.

The Tonic No. 2, as given in the Appendix should be used if there is much debility.

359. INFLAMMATION OF THE LUNGS (Pneumonia). INFLAMMATION OF THE SEROUS MEMBRANE INVESTING THE LUNGS, (Pleurisy,) SOMETIMES CALLED LUNG FEVER.

> Hyssop among all purging herbs is best,
> And frees from phlegm the overburdened chest.
> When cooked with honey 'tis esteemed the chief
> Of balms to give the lungs complete relief
> Its use, by some, is said to give the face
> The highest character of human grace.

In Pneumonia the right lung is more liable to be affected than the left, the lower part oftener than the upper, (in consumption the upper part oftenest). In Pleuritis, the lining membrane of the chest is the part affected; but the patient generally complains of a "pain in the side." As these diseases are of a like character, and often coexist, the same general treatment will answer for both.

THE SYMPTOMS.

Are those of Fever and Inflammation. At the commencement there is a chill, a hard wiry quick pulse, hot and fevered surface and headache, if there is *much congestion* the pulse is variable, and the surface will be cold and clammy instead of hot. When the substance of the lungs is the part affected the pain is dull aching and heavy, with a sense of tightness across the chest, there is no moisture about the nostrils, etc., in fact the secretions are "locked up;" the breathing is very much hurried; frequent hacking cough, with scanty tough iron-rust colored expectoration and sometimes it is streaked with blood; with either an inflammatory or typhoid tongue.

If the pains are *sharp lancenating* and *stabbing* with a

catching of the breath, cough short and painful, and the expectoration of a glary mucous character, it is Pleurisy. Read Scc. 325 and 295.

TREATMENT.

Give (Acon. and Bell.) alternated every half hour with (Bry. and Tartar emet.) and if there is nausea, add Ipec. A vapor or sponge bath will materially facilitate the cure.

For the cough, give (Bry., Phos. and Caust.)

For restlessness, give Atropia (3,) or Cina.

Or, give the sweating pills or powders. See Appendix.

Or, as soon as the active inflammation is subdued, give the Lung Balsam No. 1, of the Appendix.

Or, give the Carbonate of Ammonia mixture, (P. 225.) and a strong decoction or tincture of Pleurisy root.

Hot fomentations, the Stimulating liniment, or the Magic ointment should be used on the chest, soles of the feet, palms of the hands and wrists.

Cathartics are important in inflammation of the lungs ; use the General Cathartic Pills ; see Appendix.

Give Elm or Flax-seed tea freely.

The diet should be plain, and composed of easily digested unstimulating food.

360. CONSUMPTION. (Phthisis Pulmonalis.)

Consumption may be either a constitutional, or local disease ; or both local and constitutional. There is the Granular, Importhumous and Tubercular varieties ; arising from various *remote* causes ; but all from the same *immediate* cause, viz : lrritation or inflammation of the lungs. It has been estimated that twenty percent of the mortality of the human race is caused by this disease. Any cause or causes which produce diseases of the skin, ulcers, cancers, a cachectic habit of the body, and a lowering of the vitality of the system may engender consumption.

SYMPTOMS.

Increased sensibility to the effects of cold, and liability to "catarrh," there is a furred tongue, cough, with more or less fever in the day-time and sweating at night, with a burning sensation in the palms of the hands and soles of the feet. As the disease advances there is a hectic blush on the cheeks, the night-sweats increase and become clammy ; there is considerable emaciation, and the pus and mucous is often streaked with blood, or there may be *bleeding* from the lungs.

As the disease progresses, the cough becomes hollow and sepulchral and very distressing, the strength fails, emaciation becomes extreme, the pulse high but not strong, colliquative diarrhea sets in, the hemorrhage is often profuse, and the patient dies suddenly ; or gradually sinks from exhaustion.

Consumptive patients are very hopeful of their final recovery ; this is one of the peculiarities of the disease.

TREATMENT.

The *thick fur covering the tongue* indicates the need of ACIDS. Give Lemons, Oranges, sweet Grape juice, Tartaric or other acids ; to be followed by the Carbonate of Ammonia mixture, (P. 225, or No. 8, on P. 170,) a spoonful three times a day.

But if the *tongue is raw and sore* ALKALIES are required: Glycerine and the Carbonate of Ammonia mixture should be given first. For it is impossible for the system to utilise either medicine or food until the conditions here indicated are corrected.

Now take Valerian, Golden Seal, Sarsaparilla root, and Horse Gentian, one ounce each, and put them in a quart of good rye whiskey; take of this a tablespoonful three times a day. At the same time use the Lung Balsam (No. 1, of the Appendix) it will control the cough.

If the patient is a woman, there will probably be a derangement of the periods; then give (Mac., Pod. and Calc. c.,) until she is regular.

In case there is *diarrhea* use one of the following powders. Sulphate of copper half a grain, and extract of opium one grain.

Or, extract of Bayberry, Camphor, and Carbonate of Ammonia, equal parts, well pulverised and mixed.

DOSE: three grains.

For the *night-sweats* give Atropia (3,) in *some* cases the (2d) may be given, but with caution.

Or, Muriatic acid thirty drops, Quinine 15 grains, Alcohol and Water half an ounce, Sage and Tansy tea half a pint. Mix.

DOSE: a spoonful every three hours; also drink freely of Sage tea.

Ipecac Ointment rubbed over the chest three times a day and heated in, is useful in this disease.

Or, boil a dozen Lemons, press out the juice, add one

pound of Sugar, half a pint of the juice of fresh Horehound, and make into a syrup.

Dose: a tablespoonful three or four times a day.

Eat two pounds of sweet Grapes three times a day; and use Cream and fresh milk as freely as possible.

The milder remedies are:

Acon., Bry., Phos., Caust., Rhus tox., Apis, Sulph., China, Aurum, Dros., Calc. c., Merc., Ars. and Iod.

ACCESSORYY MEASURES.

Bathe twice a week in tepid water, in a warm room; after bathing, rub the body and limbs with Almond, Sweet Oil, or Sweet Cream. Open air exercise should be indulged in daily; keep in a cheerful mood. Eat *all* that your stomach will digest of ham, raw eggs, roast beef, *sweet cream*, oysters, *grapes* and celery; and you will be satisfied with the result.

> Goat's milk and camels' as by all is known,
> Relieve poor mortals in consumption thrown;
> While asses' milk is deemed far more nutritious,
> And, e'en beyond all cows' or sheeps' officious.
> But should a fever in the system riot,
> Or headache, let the patient shun this diet,
> Cows' milk gives wonted heat to every part,
> And quickly dissipates the acrid smart
> Of tainted humors, with a soothing art.

361. BLEEDING FROM THE LUNGS. (Hœmoptysis.) BLEEDING FROM THE STOMACH. (Hœmatemesis.)

When the bleeding is from the lungs, it is of a bright red color, and often frothy; but when from the stomach, it is dark, and sometimes clotted, or mixed with food. The treatment for both is practically the same.

TREATMENT.

Any of the following remedies are quite effectual.

1. Give (Acon., Bell. and Bry.,) alternated every half hour, or every two or three hours, (owing to the urgency of the case) with the tinc. of Hamamelis in from one to twenty drop doses.

If there is vomiting, or if from an injury, (Arn., and Ipec.) After hemorrhage with feeble pulse, fainting, cold hands and feet, anguish, palpitation of the heart, burning heat and thirst; give China or Ars.

2. Or, Calcined Deers' Horn. See Appendix.

3. Or, from four to six drops on sugar, of oil of Erigeron Canadensis (Canada Fleabane, Colts'-tail), to be given as often as the urgency of the case requires.

4. Or, oils of Cinnamon and Erigeron one dram each, alcohol two ounces. Mix.

Dose: from ten drops to a teaspoonful every ten, fifteen, or thirty minutes.

5. Or, put Sulphuric acid two and a half drams in a mortar, and slowly add oil of Turpentine one dram, stirring constantly, add slowly Alcohol one dram, continue to stir until the fumes cease to arise ; put it in a bottle with a glass stopper.

Dose : put a spoonful of sugar in a cup, on which put half a teaspoonful of the mixture, stir in water until the cup is nearly full ; then drink at a draft ; repeat every hour until the bleeding is controlled.

Or, Muriatic Acid half a dram, Carbonate of Soda two

drams, tinc. of Lupuline half an ounce, Water five and a half ounces. Mix.

DOSE: a tablespoonful three times a day.

Or, Glycerine given in one or two teaspoonful doses every two or three hours.

Or, large doses of Capsicum and Bayberry every ten minutes. It is helpful to use as a drink Gum or Elm water.

Or, Salt, in teaspoonful doses.

ACCESSORY MEASURES.

Loosen the clothing about the person; let him lie down with the head elevated and keep *perfectly quiet ;* give him fresh cool air; and ice cold Lemon juice as a drink. Bathe the feet in hot water; and the chest with the Stimulating Liniment as well.

The smoke or vapor from Frankincense, Rosin, Myrrh or any of the balsams will contribute towards a prevention of a recurrence of the trouble.

Avoid *all* excesses, and keep the secretions and excretions in a normal condition.

362. HEART DISEASE—PALPITATION OF THE HEART. ETC.

"An irregular beating and fluttering motion of the heart, very frequently attends dyspepsia, sick-headache, neuralgia, asthma, consumption, hysteria, gout, constitutional debility, and in persons of nervous temperament, it is frequently occasioned by any slight derangement of the system. The using of strong coffee, tobacco, an oppressed condition of the stomach from food that is hard of digestion, grief, anxiety, or any sudden emotion of the mind, will, in some, occasion irregular throbbing and fluttering motions of the heart."

The beating of the heart is sometimes so violent, that it may readily be perceived by the eye, or even the ear of the by-stander. The affection is usually attended with a difficulty of breathing, paleness of the countenance, and a purple color of the lips. Estrangement or seperation from *loved ones* sometimes causes a "broken heart."

These symptoms are, however, not always the effects of functional derangement; but may result from organic disease of the heart or its large vessels. When this is the case, it is generally very difficult to cure. Heart disease is accountable for very few of the sudden deaths which are said to arise from that cause. To ascertain the real origin of the sudden deaths, an experiment was tried and reported to a scientific congress at Strasburg. Sixty-six cases of sudden death were made the subject of a thorough post mortem examination. In these cases only two were found who died from diseases of the heart. Nine out of sixty-six had died of apoplexy, while there were forty-six cases of congestion of the lungs; that is, the lungs were so full of blood that they could not work, not being room enough for a sufficient amount of air to support life.

CAUSE OF CONGESTION.

Cold feet, tight clothing, costive bowels, sitting until chilled after being warmed with labor or a rapid walk, going too suddenly from a close room into the air, especially after speaking, too hasty walking, or running to catch a train

These causes of sudden death being known, an avoidance of them may serve to lengthen many valuable lives, which would otherwise be lost under the verdict of the heart complaint. That disease is supposed to be inevitable and incurable, hence many may not take the pains they would to avoid sudden death, if they knew it lay in their power.

TREATMENT.

If possible, avoid or remove any known cause. Then give (Lach., Sepia and Cactus) morning and night. If the patient is plethoric, with hot and flushed face, or congestion; give (Acon., Bell. and Gelsem.); From grief (Ignatia and Nux.) From passion, joy or fright (Cham., Coff., Verat.) From deranged period and hysterical symptoms; give (Mac Puls., and Caul.,) or the strong tinc. of Mac. and Caul., in ten drop doses.

Give a dose every fifteen or twenty minutes until relieved; then two or three times a day.

Or, a continuous and mild application of Electricity is sometimes very effective.

362. ASTHMA.

Asthma is an affection of the breathing apparatus, characterised by great difficulty of respiration, tightness across the chest, and a pressing sense of suffocation. The disease generally occurs in paroxysms, which come on in the night, and the patient is often compelled to leave his bed, and seek the fresh air. As the paroxysm comes on, the breathing becomes more difficult, laborious, and *wheezing;* the patient grows restless, his lungs seem to swell so as to fill his chest; the pulse is generally full and quick, the face often bloated, and the veins of the head and neck turgid; there is often a short dry cough, which is sometimes attended with slight expectoration, or raising of phlegm. The extremities are cold, pulse small, and often there is *palpitation* of the heart. The paroxysms are intermittent.

CAUSES.

Asthma may be owing to constriction of the windpipe, or bronchial tubes, or it may affect the substance of the lungs: And may be caused by inhalation of irritating vapors, dust or smoke, a cold, a moist atmosphere, sudden changes of temperature, *violent* mental emotions or excessive exertion, MALARIA, or a *depraved* condition of the system.

TREATMENT.

To Relieve.

First relieve the paroxysms by inhalation of the smoke from the following: either burned and drawn through a funnel or smoked in a common clay pipe. The leaves of Stramonium, Mullen and Lobelia equal parts; saturate the leaves with compound tinc of Benzoin in which has been dissolved Salt Petre, two ounces to the pint; dry and add pulverised Sanguinaria, Sumach berries or bark, and Skunk Cabbage equal parts; two parts of the first, to one of the last.

Bathe the throat and neck three or four times a day with the compound Liniment of Stillingia.

Or, take of the compound tincture of High-Cranberry-bark, (Viburnum Opulus,) from ten drops to one ounce, two or three times a day.

Or, saturate blotting paper with the following: dissolve Salt-petre one-fourth of a pound in a pint of water; add one ounce each of Fluid Extract of Stramonium and Benzoine; after it is well filled with the mixture, dry, then cut the paper into strips one-eighth of an inch wide, burn and inhale the smoke until relieved.

To Cure.

Fluid Extract of Turkey-corn one ounce, Fluid Extract of Podophyllin two drams, Fluid Extract of Leptandrin four drams, Hydrastes c. two drams, Iodide of Potassium one ounce, Salicine two drams, Water one and a half pints; mix. And give before each meal a dessert spoonful; gradually increase the dose for a week, then decrease for a week, interrupt for *a day;* then repeat.

Also use the Salicylic acid mixture, (P. 225) two or three times a week.

Or, drink freely of a decoction made from the root of the Asthma Plant: (It grows in the unbroken prairie, and resembles Black Cohosh, only the flowers are yellow, which gives the terminal, wand-like stem, that rises sometimes six feet from the ground, the appearance of corn tassel.)

Be careful and avoid as far as possible all unnecessary exposure; especially damp rooms and dust.

A dry or mountainous country is best suited for persons suffering from asthma. Such is Northern Kansas, Southern Nebraska, Colorado and other parts of the "Great West."

Theoretically the above may not please, practically it will.

DISEASES OF THE ORGANS OF DIGESTION.

363. DYSPEPSIA. (Indigestion.) BLUES. (Hypochondriasis.)

"Dyspepsia, strictly speaking, is only a symptom of disease, as it only implies difficulty of digestion, which is an attendant or effect of nearly all diseases. The stomach, which is the principal organ of digestion, is, as it were, the centre of sympathies, and thus may be reached by a great variety of morbific influences. In chronic inflammation of the mucous surface of the stomach and upper part of the intestines, we witness a train of symptoms that come the nearest filling the description generally given of dyspepsia. There is a pain and burning in the stomach, with nausea and occasional vomiting; nidorous eructations; raising of sour food and fluids of a disagreeable acrid taste; heartburn; waterbrash; constipation of the bowels; foulness of the tongue; flatulency; palpitation; epigastric pulsation; general debility; dizziness, and sometimes fainting; lassitude and low spirits. The appetite is generally poor, but sometimes voracious. A full meal, and even sometimes a very small one, will lay heavily on the stomach, especially if the things eaten are hard of digestion, such as warm, heavy bread, fat meats, and certain kinds of vegetables.

When the disease extends through the upper part of the intestines, there is generally considerable derangement among the hepatic and biliary organs.

It must be observed that the mucous surfaces, from the

nature of their use, cannot be favored with the protection of an epidermis, as is nearly every other surface of the entire system ; but the nerves and vessels with which these organs are so abundantly supplied, are invested only with a membrane of the most delicate structure. When, in view of this extreme delicacy, it is considered that the human stomach is the devoted receptacle of all that is called for by our depraved appetites, which are only governed by the capriciousness of an imagination stimulated by the corrupt fashions of our day ; and what is worse, fashion, by its magic influence, has led us to impose upon our stomachs services so unnatural and incongruous, as to make of them a kind of portable apothecary's shop, or dispensary, where drugs of every kind are kept, even without bottles or envelops, for distribution ; it will not then seem strange that inflammation of the mucous membrane of this organ should be of such frequent occurrence.

There can be but little doubt that dyspepsia, in the large majority of cases, is produced by inflammation of the gastroenteric mucous membrane. Dr. Eberle remarks: "The worst forms of dyspepsia, and all that host of inveterate gastric and billious disorders of which so much is heard, and the true nature of which is so often misunderstood, are in nine cases out of ten, the consequence of a chronic inflammatory condition of the lining membrane of the stomach."

THE "BLUES" should have the same treatment as Dyspepsia. There is languor and irresolution. The patient is particularly attentive to the state of his health ; any unusual sensation is considered as a premonitory symptom of approaching disease or death, etc.

TREATMENT.

The patient should study and practice the "Laws of Health," as given under that head, Sec. 290 : and throw physic to the dogs. The mind should be diverted entirely from the stomach ; forgetting as far as may be that he has a stomach. *Traveling*, or *having the mind engrosed* in some *speculation* is beneficial to many. Moral and religious conversation, is beneficial, if in this way the mind can be diverted from personal ailments.

Then use the following without limit.

LAUGH.

"Would you be wise ?
Take my advice :
Throw off your ills and laugh :
Banish your sighs
And dry your eyes,
Forget your cares—aud laugh.

Should friends you have trusted with confidence turn
From your proffers of friendship, you have but to learn
That if you should drain Despair's cup at one quaff,
It will not reinstate you as quick as one laugh.

Misfortunes may trip you, calamities dire
May o'ertake you, surround you with flood or with fire,
Consume all your riches as lightly as chaff—
But the victory's yours if yon sit down and laugh.

The longer you brood o'er your sorrows and woes,
More gigantic they seem, while fantastical foes
Will loom up before you, and, like the barbed gaff,
Will transfix you, unless you ignore them and laugh.

When good *business* friends stop you out on the street,
And will squeeze your hand, talking and smiling *so* sweet,
Caressing you softly an hour and a half,
Just say nothing, but look in their fond eyes and laugh.

Your true (?) friend may whisper, ' It's shameful, but true,
These old gossips do talk like the mischief of you ;'
Just make up your mind he's a natural calf,
Disgust him exceedingly by a good hearty laugh.

Our two eyes were given us that we might see
Our own way through the world; 'tis our own fault if we,
Preferring the guidance of every one's eye,
Should be forced—not to laugh—but eternally cry !' "

When medicine is used, the following will be found equal to any.

(Nux., Pod. and China) ; (Sulph., and Hydras.); to be used morning and night ; the first in the morning and the second at night.

If there is constipation of the bowels; use an injection of tepid water with a spoonful of Salt in it.

Or, eat nothing but sweet Grapes or Grape-juice, Oranges, Celery, *Roast Beef,* Cranberry sauce and *Brown* Bread.

Or, Capsicum, Mustard and Carbonate of Ammonia equal parts ; made into common sized pills with beef's Gall, and one taken at night will regulate the bowels and assist digestion. Also teaspoonful doses of the Yellow Ladies' Slipper.

Or, Apocynin ten grains, Strychnia one grain, Hydrastin twenty grains, Lupulin forty grains, Extract of Cinchona a sufficient amount to make a mass. Mix intimately and make into forty-eight pills.

DOSE: one before each meal.

Or, Tonic No. 3, of the Appendix.
Or, Electricity or Magnetism.

> Cheese is a surly and capricious elf,
> Digesting every substance but itself,
> Yet those in health their hunger can appease,
> With nothing better than plain bread and cheese ;
> But poor *dyspeptics* ever must beware,
> How they *mix bread* with this deceitful fare.

364. VOMITING.

Vomiting is caused by irritation or inflammation of the mucous membrane of the stomach, or by nervous sympathy with other parts, diseases of the *brain* or other nervous centers ; pregnancy ; hysteria ; obstruction of the bowels ; too much or improper food ; *medical gluttony ;* cancer of the stomach ; bilious and eruptive fevers ; starving or excesses of any kind.

If vomiting gives relief it should be encouraged, it is a favorable indication ; but if the symptoms preceding vomiting are increased it portends danger.

TREATMENT.

If from debility of the stomach or biliousness (Nux and Pod). If from rich food, or over-loading the stomach (Puls., and Ipec.) If there is diarrhæa and a burning of the stomach (Ars., and Lept.)

Or, give (Verat. a., Arsen., Cupr.,) after vomiting.

Or, Dioscorein, ½ grain doses, in a teaspoonful of Peach-leaf tea.

Or, if from pregnant sickness, give ten grains of Oxalate of Cerium after each fit of vomiting; if there is much irritation, dissolve two or three grains of Bromide of Potash in water, and take at a dose.

For further remedies see Cholera Morbus, Cholera and Cholera Infantum.

365. COLIC. (Enteralgia.)

It may be known from inflammation of the bowels, by the absence of fever, and from the fact that pressure gives relief; otherwise the symptoms are too well known to require a description here.

TREATMENT.

It may be divided into three varieties, viz: Wind Colic, Painter's Colic and Bilious Colic.

Painter's Colic, which arises from lead poison, should be treated as follows.

Boil two drams of powdered Alum in a pint of Milk, then strain and give a wine glass full every three hours until it acts as a physic : in some cases a quarter of a grain of Morphine may be given for temporary relief; or Curara or Morphine may be injected subcutaneously near the most painful parts : otherwise the treatment should be the same as for the other classes of colic.

> Four ills from long-imprisoned flatus flow,
> Convulsions, *colics* dropsies, vertigo;
> The truth of this the thing itself doth show.

Wind colic should receive the same treatment as that given for colic of Infants (P. 190).

For Bilious Colic, give Dioscorein in half grain doses every hour.

Or, give from ten drops to a teaspoonful of the compound tincture of Cajeput every fifteen or twenty minutes until relieved.

Or, give Peppermint in Alum Whey.

Or, tinc. of Camphor and Sulphuric Ether equal parts.

DOSE: from one drop to a teaspoonful every twenty minutes until relief is had.

Or, give the Calomel Powders. (See Cholera.)

In obstinate constipation with much pain; use an enema composed of Water one quart, Beef's Gall a spoonful. Also give the "Anti Bilious Physic." Otherwise give the same treatment as is given for Cholera or Cholera Morbus.

366. CHOLERA AND CHOLERA MORBUS.

> A stated diet, as it is well known,
> Of physic is the strongest corner stone.
> By means of which, if you can naught impart,
> Relief or cure, vain is your Healing Art.

While these are different diseases, yet the same medicines which cure Cholera, will promptly cure Cholera Morbus.

In Cholera Morbus, the discharge consists of what has been taken as food, and is ejected again, more or less changed; while in Cholera, the discharges, after the first two or three, consist only of a fluid like rice water, which is, in fact, the serum of the blood. The coldness, blueness, cramps, and other symptoms are much more violent in Cholera than in Cholera Morbus.

PRECAUTIONARY MEASURES.

During the prevalence of Cholera, care should be taken that no decayed vegetable or animal substances be permitted about the house or yard. Keep the cellar dry, whitewashed, and the yard and kitchen free from slops. Copperas and Chloride of Lime should be used freely about the privy and sink. Eat a simple but nutritious diet: as beef, mutton, broiled bacon, rice, farina, crackers, stale bread, etc. Avoid everything that is hard of digestion; for drink use pure water. Excesses of every kind, mental as well as physical, should be carefully guarded against; make no *sudden changes*, the usual habits should be maintained, so far as they are not detrimental to health. Eat your meals regularly, and go to bed early.

Avoid all irregularities, night air, late suppers and late hours. Take exercise, but avoid fatigue. Take food *before going out* in the morning, and maintain a cheerful, even tone of mind. Let your clothing be adapted to the temperature of the weather, employing rather more than usual; and use flannel, especially about the abdomen, with a view of guarding against sudden changes. Do not neglect slight indispositions, especially of the stomach and bowels, but attend to them at once, as they are often the precursors of the disease itself.

TREATMENT.

The first remedy is almost a specific in uncomplicated Cholera, viz: Boracic Acid one dram, Soda bicarb., twenty grains; mix and divide into six powders; one to be given every two hours.

Or, Oils of Cajeput, Aniseed, Peppermint, and Cloves, of

each half an ounce ; Tinc. of Camphor two ounces ; mix.

DOSE : from twenty drops to a teaspoonful every ten or fifteen minutes until the spasmodic action is subdued, and reaction established.

Or, Oils of Fireweed and Erigeron, equal parts ; Mix.

DOSE : one to ten drops on sugar, as often as needed.

Also use injections as in Dysentery of children.

Or, Laudanum, Tinc. of Camphor, Tinc. of Rhubarb, and Ipecac, equal parts ; mix and give half a teaspoonful, at each movement of the bowels. If the diarrhea is not checked, increase the dose at each movement until they *are* checked.

Or, give alternately Nos. 2 and 5, P. 200.

Or, take Calomel one grain, Camphor ten grains, Ipecac and Morphine two grains each, thoroughly mix, and divide into twelve powders.

DOSE : one powder every half hour at first, then extend the time to three hours.

After it is under control give the Carbonate of Ammonia and Salicylic acid mixture, P. 225.

Or, of the milder remedies,

Give (Arsen., Verat. a., and Cupr.,) ; to be given every fifteen minutes at first, and the time between doses extended as the cure progresses.

An occasional dose of Sulphur, and wearing a Sulphur *Pad* over the stomach or under the soles of the feet, is a good preventative measure.

Hot Oats, or Corn, put in sacks and the patient literally packed in them, and the hot bath thoroughly used, is often of benefit. The Bathing Drops, or Stillingia Liniment may be applied with much friction ; cover well and sweat.

The Carbonate of Ammonia mixture should be given to the convalescent.

Keep perfectly quiet, and let the diet be bland but nourishing.

367. JAUNDICE. (Icterus.)

"Jaundice is occasioned by an obstruction of the course of the bile, which is consequently taken up by absorption, and thus carried into the circulation ; and hence, by its diffusion into the capillary vessels, stains the skin, eyes, mouth, throat, urine, etc., (being the only places in which it can be seen, although affecting the whole body just as much,) with a deep yellow.

This obstruction of the bile may arise from the presence of calculi in the gall-bladder, from inflammation of its duct or even the liver itself. The obstruction may also be produced by constipation of the bowels, external injuries, and mental emotions.

The attending symptoms are loss of appetite, dyspepsia, vertigo, nausea and perhaps vomiting, flatulency, and great dejection of spirits."

TREATMENT.

The treatment, in this complaint, should be commenced by taking a pill composed one quarter of a grain each of Podophyllin, Leptandrin, Mercurius v., and the extract of Dandelion, every two hours until a cathartic effect is produced. This compound seems to have a specific action on the glandular system, and will generally reach the liver, stimulating it to healthy action.

The yelk of Eggs has been highly esteemed as a remedy in Jaundice. It has to be taken raw on an empty stomach.

Or, a gill of Sweet Oil three times a day.

Or, ten grains of Salicilic Acid in Soot tea, three time a day.

Eat acid fruits.

> Pain of right side and loaded tongue show bile,
> Vomitings, vigils, ears that ring awhile.
> Thirst, nausea, looseness, colics painful smart;
> Poor appetite and biting cramps of heart.
> Bile spreads throughout, its bitter taste of gall,
> Burns, and gives rise to fiery dreams in all:
> With yellow tinted skin like leaves in early fall.

368. LIVER COMPLAINT—INFLAMMATION OF THE LIVER. (Hepatites.)

SYMPTOMS.

An acute attack is ushered in by many of the symptoms common to fever. As rigors, hurried pulse, furred tongue, thirst, hot dry skin, cough, headache, and constipation; *pain extending to the top of the right shoulder*, tenderness of the right side, increased by strong breathing; the breathing is thick and heavy; sometimes there is jaundice and nausea. In *chronic* inflammation of the liver, the liver may be enlarged or diminished in size; there is sallowness of complexion, languor, lowness of spirits, and a dread of some impending evil.

CAUSE.

Any cause that hinders the elimination of the waste or

worn-out matter of the system, or lowers the vitality, or causes an excess or accumulation of the *alkaline* element in the system.

TREATMENT.

When acute with feverish symptoms, shooting pains, and a sallow yellowish tinge ; (Acon., Merc., and Bry.,) every hour. When the active fever symptoms are controled : then give (Nux,, China, Pod. and Sulph.,) three times a day.

Or, Electricity and Cold Bathing.

Or, use the Liver Pad of the Appendix, and give the Alterative Tonic.

Or, Sanguinarin, Leptandrin, Podophyllin and Horse Gentian, half a grain each three times a day.

Followed by the Salic lic mixture each alternate day.

Or, the same as for jaundice.

Or, from five to ten grain doses of Sulphate of Manganese in water, to which add ten grains of Citrate of Magnesia.

BILIOUSNESS, HEART BURN, FLATULENCY, ETC.

Commonly called "Bilious Attacks," are generally supposed to be due to derangements of the liver, are almost always symptoms arising from Indigestion, or Dyspepsia. When there is *no inflammation of the liver*, they should be treated as Dyspesia.

369. DIARRHEA—PURGING

In addition to the treatment given under he head of " Diseases of Childhood.' the following will be found useful.

TREATMENT.

Use the Diarrhea Mixture of the Appendix.

Or, in Typhoid diarrhea there is nothing better than the following : Luca Oil thirty drops, Oil of Turpentine fifteen drops, made up with mucilage, or the yelk of an egg ; to be given at a dose and repeated as often as required.

370. DYSENTERY—BLOODY FLUX.

In addition to the following, see what is said on the same subject under "Diseases of Childhood."

Give the White Liquid Physic of the American Dispensatory : a spoonful every hour until it physics, then every three hours.

Or, Compound Tinc. of Rhubarb in teaspoonful doses.

Or, the Dysentery Pills of the Appendix, which are almost a specific. Or, the Salicylic Acid mixture, P. 225.

Or, No. 8, P. 170.

371. CONSTIPATION, COSTIVE BOWELS.

> At least three times in every fleeting day,
> Some tribute to the renal functions pay,
> And once or twice all alvine calls obey.
> Such ones as dare her warning voice neglect,
> May soon some unrelenting ill expect.

Constipation of the bowels, although a common affection, seldom occurs without the existence of some other derangement. It most commonly attends inflammatory affections of the liver, brain, stomach, bowels, and it is one of the marked symptoms of dyspepsia.

Among the most common symptoms, collateral with constipation, are sickness of the stomach, want of appetite, flatulency, headache, fever. general dullness, and dejection of spirits. TREATMENT.

Constipation will be found, generally, to yield to the appropriate treatment of those diseases that give rise to this complaint. But in confirmed cases it becomes necessary to do something that is more directly calculated to relieve this state of the bowels. For this purpose, whatever can be effected by way of dieting, should be done; for it is not a good practice to use much cathartic medicine for this purpose. Most articles of this class are apt to leave the bowels in a condition little better than they find them. Moreover, purgative medicines, if much used, are almost certain to injure the digestive organs, and thus they ultimately very much increase the difficulty. In mild cases, the patient will generally be able to keep the bowels in a proper condition by the use of ripe, jucy fruit, such as peaches, &c. The Graham or unbolted bread is also eminently adapted to costive habits. With some, potatoes, if used twice or thrice a day as food, will keep the bowels in a free and healthy condition. Pepper sauce, used at the table, will be found serviceable to those troubled with weak digestive organs."

In obstinate cases a more active course will sometimes be required. The bowels should, in the first place, be moved by injections made of beef's Gall and Glycerine, a spoonful each in a pint of Water.

(Nux and China) alternated with (Sulph. and Hydras.) the first in the morning and the second at night.

Or, the same as for Liver Complaint.

Or, Tinc. of Cascara Sagrada and Glycerine, equal parts. Dose: a teaspoonful at night.

DOSE OF HOMEOPATHIC MEDICINES.

"In determinging the quantity and strength of doses several circumstances should be considered: such as age, sex, habits, nature of the disease etc. As a general rule, without reference to individual peculiarities, the following may be stated as the proper dose in domestic practice: FOR AN ADULT, ONE DROP OF TINCTURE, FOUR GLOBULES, OR ONE GRAIN OF TRITURATION; FOR A CHILD, ABOUT ONE-HALF THE QUANTITY; FOR AN INFANT, ABOUT ONE-FOURTH.

A drop is easily divided into two doses, by mixing it with two spoonfuls of water, and giving one spoonful for a dose.

REPETION OF DOSES.

In the repetition of doses we must be guided by the acute or chronic character of the malady, the urgency and danger of the symptoms, and the *effects* produced by the medicines. In violent and acute diseases, such as cholera, croup, pleurites, convulsions, etc., the remedies may be repeated every ten, fifteen or twenty minutes. In less urgent cases of acute disease, the remedy may be repeated every two, three, or four hours. In all cases when *improvement* takes place, the medicines should be taken less frequently, and gradually relinquished." (E. H. Ruddock, M. D.)

TIME.

As a general rule the *most* suitable time for taking the medicines, is about half an hour *after*, or before a meal; in the morning on rising, and at bed time; but if the exigency of the case requires it; at any time. The powders or globules may be either taken dry, or dissolved in a spoonful of water for each dose, and given in spoonful doses; the

Tinctures are given in the same way, and are more active and reliable in acute diseases.

The list of *medicines*, with abreviated and full name ; dilutions or triturations recommended for each one ; together with the diseases in which they are especially indicated ; are arranged in alphabetical order .

MISCELLANEOUS DISEASES.

373. AFFECTIONS OF THE URINARY AND GENITAL ORGANS.

INFLAMMATION OF THE KIDNEYS. (Nephritis.)

Diseases of the kidneys like most other diseases may be either acute or chronic. In the acute form the pain is more severe, and attended by the usual symptoms of fever. The usual symptoms are : A deep-seated pain and weakness in the small of the back ; frequent desire to urinate, yet the discharge is scanty, mudy, various colored, with a puslike brickdust or bloody sediment ; with an irregular flow ; and sometimes there is vomiting, and numbness along the thigh of the affected side. Treatment. See division 6, below.

2. INFLAMMATION OF THE BLADDER. (Cystitis.)

If the internal or lining membrane is affected, there is a *burning* pain ; if the outer covering, *piercing*. Excepting the location (this being low down in front) the symptoms are very similar to those attending Disease of the kidneys.

Treatment. See division 6, below.

3. GRAVEL, STONE IN THE BLADDER. (Calculus.) AND RENAL CALCULI (Gravel in the Kidneys or Ureters.)

If in the bladder, the attacks of pain in the region of the bladder are sudden, very acute and severe ; the thighs are affected by pain and numbness. the urine stops suddenly, increasing the pains, and there is pain in the end of the penis. The pains come on in paroxyisms until the calculi are discharged.

If the attack is from gravel in the kidneys; the pains are of a similiar character; but located near the loins. Treatment. See division 6, below.

4. RETENTION OF URINE. (Strangury.)—INCONTINENCE OF URINE. (Enuresis.)—DISCOLORED, SCANTY OR PAINFUL URINATION.—INFLAMMATION OF THE URETHRA OR CONTIGUOUS PARTS.—BLEEDING.

5. The tissues forming the organs affected by the above diseases being largely the same ; it follows that the treatment should be nearly or quite the same in all: even *gall-stone*, and *stone* in the *bladder* are most successfully treated by that class of medicines which controls the *inflammation* of the bladder, liver, kidneys, etc.

TREATMENT.

6. Take Oleo-Resin Eupurpurin and Salicylic Acid one dram each, and let it be rubbed up with Prussiate of iron, until it has sufficient consistency to form pills; and divide into sixty pills.

DOSE : one or two three times a day : at the same time keep the bowels free with the G. C. Pills of the Appendix.

Or, (Bell., Canth., and Merc.,) alternated with (Puls., Lycop., and Sasapar.,) three times a day.

(Or, Apis, Arsenicum, and Bry.) the same.

Or, Arseniate of Gold (5th),

If very obstinate vapor or hot baths should be used, and at the same time from one-sixth to one quarter of a grain of Morphine given ; or a dose of Atropia (3).

Also see Diseases of Childhood.

> Cherries you'll find are of benign intent,
> They purge the stomach, and the *stone* prevent,
> The blood throughout in healthy tone augment.

374. TAPE WORM, (To Remove.)

Take bark of the root of Pomegranate and Root of the Male Fern, boil two ounces each in two quarts of water, strain and boil it down to one quart. Dose : A wine glassful every two hours until all is taken ; follow then with a brisk cathartic. Or Kamala, the same.

3. Or, Kousso: A handful of the flowers macerated in three gills of lukewarm water for fifteen minutes. To be taken on an empty stomach, preceded and followed by lemon juice; low diet for two or three days previous, and an occasional physic, to be followed by a gentle cathartic.

4. Or, Elm bark chewed and swallowed.

5. Or, Pumkin seeds freely used.

Each one of these is reputed to expel the Tape worm.

Kamala. Dose from one to three drams. If it fails to operate, give a dose of oil.

375. PILES (Hœmorrhoids).

This common and well-known disease always depends upon a congested condition of the venous circulation of the abdomen and its immediate connections. The final result is engorgement, distention, and inflammation of the vessels distributed over the rectum, causing engorgement, tumors and bleeding of the parts.

Any treatment of tnis disease to be successful, must first remove the abdominal congestion, then the local treatment will not be difficult.

TREATMENT.

1. Give (Ars. Acon. and Hamamelis) three times a day for two days, then (Hamam., Nux., and Sulph.,) for the same length of time ; alternate these until cured. As a local application use any of the following : The Hawk Eye Salve ; Stramonium and Hamamelis Ointments equal parts ; Buckeye Ointment ; a leaf of Tobacco bound on the parts ; Cutis or the Carbolic Acid Ointments ; or, an ointment made by simmering Hamamelis, Stramonium and White Oak Bark in the fat of *old* bacon ; or, the continued use of hot or cold Water to the parts ; or, Electricity.

2. Or, put Pine-knots in Whiskey and use a tablespoonful three times a day.

Or, a spoonful of Glycerine, five grains of Salicylic Acid, and five drops of Hamamelis ; mix and take this amount three times a day ; also anoint the parts with the same as well.

Or, use Carbolic Acid in the same manner as for Carbuncle.

376. INFLAMMATION OF THE SPLEEN. (Splenitis.)

The chronic form of the disease is a very common affection ; commonly called Ague Cake. The pain and swelling is in the left side near the spleen.

The same or like TISSUES being involved, both in the inflammation of the liver and spleen ; by changing from the right side to the left, the same symptoms and treatment answer for both.

377. PAINS IN THE LOINS. (Lumbago.)

The treatment should be, the use of Electricity, or the application of Irritating plaster ; or the Stillingia liniment : otherwise the same as for Rheumatism.

378. RUPTURE. (Hernia.)

As soon as the rupture is discovered no time should be lost in replacing it : elevate the hips ; draw the thighs up toward the body ; and by gentle manipulation with the fingers replace the protruding part. If at first the effort is not a success, then give a dose of Morphine, and a hot or vapor bath, and try again. Send for a *good* SURGEON.

> Elecampane brings joyous health to all
> Thoracic organs, whether great or small.
> To drink its juice, combined with that of rue,
> Is the best thing that *ruptured* men can do.

RHEUMATIC FEVER—RHEUMATISM,—ACUTE AND CHRONIC.

This disease sets in with general fever symptoms ; and in place of affecting the nervous tissues as does *ague*, the *fibrous* structures are the points of attack; as the knee, shoulder, elbow, ankle joints, pericardium and valves of the heart, *sheaths* of the muscles and nerves, etc. It is liable to move from one part to another : the greatest danger consists in its liability to attack the heart by metastasis from some other part. A *marked* indication of rheumatism, is the *cramped* appearance of the *sides* of the finger nails.

The parts affected are often red, swollen and very painful, as all other inflammations of these structures are, on ac-account of their unyielding nature.

TREATMENT.

1. Take Salicylic Acid one dram, Soda bicarb., and Carbonate of Ammonia fifteen grains each, Water three ounces.

Dose: from one to two tablespoonsful every three hours.

For external use, take yellow Clay, dry and pulverise it, and make a pad suited to the part affected ; the cloth next to the person should be thin ; saturate the clay with tinc. of Camphor as *strong* as it can be made, and *keep it wet* with it. Also use the vapor bath.

2. Or, take gum Guaiac, Colchicum Seed, and Nutmegs; one-fourth ounce of each ; Sulphur and Saltpetre one ounce of each ; pulverise and mix with two ounces of Molasses.

Dose : a teaspoonful every two hours until the bowels are moved; then three times a day.

Or, five grains of Prunin three times a day.

For external use ; add to an ounce of the Stillingia Liniment Sassafras oil one dram, Chloroform and Tinc. of Camphor one ounce each ; keep it well corked. Apply three times a day rubbing it thoroughly in.

Or, Electricity.

Or, treat as for ague.

The Milder Remedies.

If there is much fever (Acon., Bell., and Bry.,) then (Bry., Tart., and Rhus tox.,) every two hours. If it is a woman alternate with (Puls., Me c., and Caul.,) and Sulph., an occasional dose.

Externally ; use Hamamelis, Arnica, and Rhus tox., or Calendula liniments.

Or, Aconite and Hmamelis ; one part of the first, and four of the latter ; mix.

Live on *acid fruits*, Celery, brown bread and roast beef.

380. GOUT. (Podagra.)

Gout may be distinguished from Rheumatism, from the fact that it generally is confined in its attacks to the middle aged and old persons ; and is produced by luxurious living, or is hereditary ; it chiefly affects the *small* joints ; it is not attended with acid sweats as is rheumatism, etc.

TREATMENT.

The same as for dyspepsia and constipation.

381. DROPSY. (Anasarca.)

Especially Dropsy after Scarlet Fever : give (Apis, Arsen.,

and Bry.,) alternated with (Canth., Bell., and China ;) otherwise Arseniate of Gold (5th,) alternated every other day with Bromohydrate of Caffeine (1st.)

Or, of the more heroic remedies, any of the following will be found quite efficient : Scoparius (Common Broom,) Indian Hemp (Apocynum,) Milk-weed, Dwarf Elder, Oleo-Resin Eupurpurin (from Queen of the Meadow,) or Oil of Sun Flower Seed. Juniper berries may be used in connection with all. Use *acid drinks freely*, and every other day take grape vine ashes, a teaspoonful.

Or, No. 6, P. 268.

382. SCROFULA,—VENEREAL DISEASES (Syphilis etc.,) AND CANCER.

While Scrofula and Cancer are not generally considered contagious ; yet for domestic practice the most essential treatment is the same for all.

TREATMENT.

Take Corydalis and Hydrastis equal parts, make a strong decoction, and use it *freely* three or four times a day ; at the same time wash the ulcers or diseased parts with the same. The remedies Corydalin and Hydrastin in two grain doses may be substituted for the above if desired ; but they are not quite as effective.

At the same time the Alterative Powders, or Liver Pills should be given ; a dose three times a week.

If there is trouble in making water, take one ounce each of Glycerine and water, and half a dram each of Canthard es and Belladona. Mix; and give a teaspoonful every three hours until relieved.

Or, Arseniate of Gold or Platina (4th or 5th); at the same time treat the *ulcers* the same as for CARBUNCLE.

In Cancers the free use of Red Clover Blossoms in decoction, tincture or extract internally, has gained favor with the knowing ones; at the same time the concentrated juices of Sorrel, and Clover, with the Sulphate of Zinc should be made into a plaster and used as an escharotic on the Cancer. (In Gonorrhæ, keep the bowels free with Liver Pills; use *three* ten drop doses two days apart, of Balsam Copabia, and at the same time wash and inject the parts with the following : Tincs. Lobelia, Myrrh, Hydrastes and Balsam of Fir, equal parts. There is generally two much medication in this disease, which increases the trouble.)

Heal the sores with any of the healing salves of the Appendix.

Shoemaker's Wax applied to the sores often cures.

The Chlorides of Zinc, Bromine, Gold, or Antimony, may be substituted for the Sulphate of Zinc.

As a last resort, like amputating a limb; use the following *not* very safe remedy, viz : Half an ounce of Cream of Tarter, two drams of Mercury (quicksilver), one ounce of Sulphur, two drams of Saltpeter. Put the Mercury and Sulphur in a mortar and rub them together for a *long* time; then add and mix all with a pint of Molasses.

DOSE : a tablespoonful at night.

Poultices of Figs; Carrots; Elder; Clay; etc.; are sometimes useful.

Other remedies. (Baryt. c., Silic., and Lach.,) alternated

with Arseniate of Gold (5th,) and Alstonia Constricta (2d).

No pork should be used; but plenty of roast beef and fruit. As far as possible use only the internal remedies.

383. DRUNKENNESS. (*To Cure.*)

> New wines inflame the breast, the reins excite,
> Injure the brain, and have a burning might.
> Dark wines are the quickest to intoxicate,
> To burn, destroy, and constipate.
> From beer *gross humors* and great strength will start,
> And *sizy blood* be formed in every part.
> It spurts the reins and flesh augments in all,
> The bowels frees and e'en *distends* withal.

1. Take a quart of Whiskey and put a *fresh unwashed Eel* Skin in it, and shake them together for five or ten minutes: let the patient drink a few drams of this, in the ordinary way, and he will think that he is going to die; but he will recover; and afterwards he will not be able to endure the smell or taste of Whiskey.

2. Now take Blue Scullcap root, Babyberry bark, Hemlock bark, and Ginger root one ounce each; Yellow root, Cayenne Pepper, and Cloves two drams each: let all be well pulverised and mixed. To half a teaspoonful of this put a teacupful of hot water, and drink half on rising in the morning and the balance just before eating; so use it for a week or two.

3. Or, take *alcoholic extracts* of Yellow Lady's Slipper, Blue Scullcap, and fresh Pernvian bark equal parts; and make into three grain pills. At first give from two to four

pills every three hours; gradually decrease the number as improvement progresses.

4. In DELERIUM TREMENS, an infusion of equal parts of the *fresh* roots of Blue Skull-cap and Yellow Lady's Slipper is almost a specific.

DOSE: from a tablespoonful to a teacup half full as occasion may require.

384. MAD-DOG BITE. (Hydrophobia—*To Cure.*)

Wash the wound *perfectly clean* with Vinegar of Lobelia; then dry it. Now pour into the wound a few drops of Muriatic or Carbolic Acid to destroy the poison of the dog's saliva. After the Acid has had its full effect, apply a poultice made of equal parts of Slippery Elm and Lobelia Seed (pulverised) wet with a solution of Carbolic Acid; change it three times a day until the wound is healed.

At the same time, take the following:

Take one and a half ounces of *fresh* crushed Elecampane root, and boil it in one pint of fresh Milk until reduced to half a pint, then strain and cool. Drink at one draft. The next day use No. 4, as for Delerium Tremens. The third day two ounces of Elecampane prepared the same way. The fourth day repeat the *Delerium Tremens* medicine. And the fifth repeat the same as on the third day. Eat nothing for six hours before or after taking the medicine. If used before the spasms come on there is nothing better than the above treatment. The vapor bath raised to as great a heat as can be endured, and used every other day for two weeks, has been the means of saving many from a terrible death, even after the spasms were manifest.

385. SNAKE BITE.

Treat the wounds as for Hydrophobia, or Carbuncle.

Or, apply a poultice of Blue Bell or Rattlesnake's Master root, boiled in fresh Milk and often changed.

Or, bathe the wounds with Aqua Ammonia.

Give freely of an infusion of Blue Bell or Rattle Snake's Master root.

Or, kill poison with poison, by drinking Whiskey until partially intoxicated.

Or, cut open a chicken and apply it to the wound.

> The radish, pear, thoriac, garlic, rue;
> All potent poisons will at once undo,

386. STINGS OF INSECTS.

Bruise and apply the pulp of any of the following : Plantin ; Lobelia ; Blue Violet ; or, the tinctures of the same.

Or, Aqua Ammonia ; Camphor ; Salt ; Clay ; or Carbolic Acid Ointment.

If there is much general disturbance of the system ; give (Rhus tox, and Acon.,) or, (Acon., Bell., and Apis.)

387. NERVOUS AFFECTIONS.

In treating of these disorders the Author has carefully

studied his subject in the light of science, and by the bedside of the afflicted; and the reader is assured that what is written, is after mature deliberation and consultation of the best authorities coupled with experience; and further, lest the character of the treatment should be misunderstood, the Author gives the assurance that it is *not* in any sense treated from a theological standpoint; but from a purely hygienic or medical standpoint.

The causes of Nervous affections are, 1st congenital; 2d. MENTAL ABERATIONS; 3d, IMPERFECT NUTRITION; 4th, injury of the nerve tissue; 5th, Poisons; 6, self-abuse. In any case there is but *one common immediate* cause, and through this, all strictly nervous disorders produce their effects on the sufferer. If this common central medium of force through which, and on which all these sources of causation act or pass, can by *any means* be controlled, strengthened, regulated, or be put into a condition of health and strength, it will become the controlling power in place of being itself controlled by the lesser forces; and will bring all of the others under subjection and into a state of normal or healthy action. This central force on which so much depends, is the *hidden* power which *actuates* the *mind*, the vital or soul force if you please; but it is out of our province in this place to discuss just what it is; but that it does exist is self evident; and without accepting this truth, it is certain that any and all attempts at solving the phenomena attending the action of what by medical men is called " the life or healing power of nature" in the human organism, must ever remain in the shadows of uncertainty. As a medical man searching after light and *truth* for the purpose of alleviating the bodily sufferings of my fellow men; I am compelled to accept the Bible as the only *possible way* out of the dark maze of uncertainty which like midnight gloom covers every other avenue of exit : and further for the benefit of those who have not considered this subject in the light of scientific truth ; let me say that of all the writers whose works have come under my observation ; *not one* has been found whose advice on that department of medical science which treats of the *preservation of health*, equals the laws given by Moses on the same subject ; he recognizes the central living force power as a part of man, and on this basis his laws are founded. And further, there is not in the whole range of medical literature, one author of any consideration, however *devoted* to the *physical* and *chemical* views of *life*, who does not evince the *necessity* of admitting a governing *Vital Principle* as a distinct entity, distinct from all other things in nature ; and is compelled to summon to the aid of his discussion, *a vital principle* whenever he touches upon the abstract phenomena of life. These remarks have been thus extended, that the reader might the more readily accept and use the treatment for nervous affections which is to follow:

TREATMENT.

First get en rapport or in harmony with the teachings of the Bible, and obey its laws strictly; but with great calmness; look away from yourself and your ailments ; let your thoughts be occupied with pure, good, useful and pleasing subjects ; and you will be surprised with the happy calming results : for imagination agrivates, yes, positively produces disease ; while a strong will prevents or controls it. Unprincipled quacks use this law of nature through their vile publications for the purpose of first alarming and then fleecing their victims; you have a better councilor; lean on the Bible and let them and their books severely alone. All of the passions are subject to the will, if the mind is in a healthy condition: then the nervous forces will be in a state of calmness and health. In addition

For 2d and 3d causes, read and apply what is said under the head of Dyspepsia and Blues, and give a dose of ATROPIA (3d,) or Tinc. of Lettuce at night,.

For 4th. Treat as for Neuralgia.

For 5th. Quit using the poison or exciting cause; lay aside the excessive use of Tea, Coffee, Tobacco, etc.

6th. Nervous debility, sexual weakness etc., give (Aurum, China. and Phos.,) morning and night. Or, Bromohydrate of Caffeine.

Or, Arseniate of Gold (5th,) alternated with Alstonia c. (2).

Or, Salicylate of Iron, alternated with Nux. v., (3) and China (3).

Or, No. 7. of the Appendix; also *No. 8.*

Use Acon., Bell., Mac., *Calc.*, *Cina.*, or such other remedies as *are proper* for the disease *accompanying* the nervousness.

Use medicine and DIET as given for the Blues.

See Spasms, as well.

388. NEURALGIA. (Tic Douloureux.)

(Acon., and Puls..) alternated with (Bell., and Nux. v.)

Or, (Mezer., Plant. m., and Bell.)

Sometimes it is necessary to give drop doses of the strong tincture, and bathe the parts affected with the same.

Atropia (3d,) *or* Phos., may be given at night.

Or, give the treatment as for Headache. P. 209 and 210. Also see P. 222.

Or, any of the ague medicines combined with a quarter of a grain of Morphine.

Or, the Salicylic Acid mixture.

Use Electricity and the Stillingia liniment externally.

Hot fomentations are sometimes of use.

389. HABITUAL COLD FEET. (*To Cure.*)

Take a dose of Calcara c., three times a day, and apply a *mild* current of Electricity to the upper extremities, chest, spine and loins.

390. MONTHLY IRREGULARITIES, ETC.

The various diseases arising from the irregularities incident to the Monthly Periods, give rise to a train of symptoms which *combine* in a greater or lesser degree, sympoms attending Fever, Rheumatism, Epilepsy and Neuralgia ; from the fact that the entire physical organization of woman is involved in sympathy with nature's effort to establish a healthy condition of the "monthly courses." While the symptoms are multitudinous ; yet as the same character of tissues or parts are involved in these various diseases known by different names ; (as in Fevers ;) there is a unity of diseased condition ; and there should in a large measure be a general unity of treatment ; these facts should be borne in mind ; as it *simplifies* for *domestic* practice what otherwise is a mass of complications. The *general* principles given for the treatment of *Fever, Inflammation,* Dyspepsia and Blues, should be carefully read.

SUPPRESSION OF THE MENSES. (Amenorrhea.)—PAINFUL MENSTRUATION, (Dysmenorrhea.)—PROFUSE OR TOO EARLY MENSTRUATION. (Menorrhagia.)—WHITES. [Leucorrhea.]—GREEN SICKNESS. [Chlorosis.] HYSTERICS, [Hysteria.]—BLEEDING FROM THE WOMB. [Metrorrhagia.)

As the MIND, Brain and other nerve centers; together with the Heart and circulatory system, are all in common sympathy with the reproductive organs; it is proper that the treatment for all, should be largely general in its character; while at the same time the specific or local treatment of the parts or organs involved should not be neglected. The various *names* given to these irregularities are merely for the convenience of authors of medical works: for *each and every case* must be treated on *its* distinctive symptoms or indications of disease, *regardless* of *name;* as is the case in nearly all diseases.

TREATMENT.

The Willow's blossoms drank in water cools
Consuming love, and fierce excitement schools,
And all productive power thus over-rules.
The leek will all young women fruitful make,
Who of its substance constantly partake.
Of all the cunning drafts that you can brew,
The *best* is Sage, combined with graceful Rue.
Let Rose-leaves be into this mixture brought,
And love's desires will quickly come to naught.

If a feverish condition exists, give a few doses of [Acon., Bell., and Bry.,] and bathe the feet in hot water on going to bed. Follow this by [Mac., Caulo., and Puls.]

Or, [Apis, Sepia and Puls.,] alternated every one, two or three hours with (Coccul., Plant. m., and Chamomilla.) Sulphur at night; and if there is much prostration; or swelling of the face, feet or ankles; give [China, Pod., and Rhns tox.,] or Arsenicum each morning. If the pains are of a wandering rheumatic character, then give No. 1, P. 272.

When lasting too long, or too profuse, [Calc c., Cocculus, and Ipecac,] or [Hamamelis., Caulo,, and Cocculus.] If injections are used, they should be *hot*, with the additional use of hot washes of Hamamelis, Hydrastis, Arnica, or Carbolic or Salicylic acids largely diluted with Water: all of the various leucorrhœal discharges, itchings, burnings, etc., may be treated the same way; only (Carbo. an., Bell., and Nux.,) or Senecio may be given in addition to the above.

STRONGER REMEDIES.

For bleeding from the Womb; after using an injection of *hot* water in which has been put a teaspoonful of Tinc. of Hamamelis and Arnica to the quart of Water : then use the treatment as given for Bleeding from the Lungs and Stomach, on P. 246 : when the hemorrhage is checked, then give the same treatment as is given for other derangements of the womb.

The most perfect treatment in deranged. monthlies, for general use, that it has been my lot to use is the following : viz : Dr. Dewee's celebrated " Compound Tincture of Guaiacum " two ounces, Tinc. Macrotin two drams. Mix.

Dose: one or two teaspoonsful three times a day, in milk or wine.

Or, Tincs. of Caulophyllin, Macrotin, Female Regulator, Trillium, and Horse Gentian, equal parts; Carbonate of Ammonia a dram to the pint. Mix.

Dose: from one to two teaspoonsful three times a day.

With either prescription; bathe the feet in hot water, or take a vapor bath if convenient.

Commence the treatment a week before the time for the catamenial discharge.

Other remedies. Horseradish and Comfrey; Helonin; Figwort; Spanish Needles.

Or, two ounces of Peach-kernels to a pint of Brandy.

Dose: of any of the first five, a wine-glass full of the infusion; of the last, a teaspoonful three times a day.

391. FALLING OF THE WOMB. [Prolapsus Uteri.]

The best practice for a *permanent cure*, is to place the positive pole of a battery on the lower part of the spine, and at the same time put the negative pole in contact with the mouth of the womb; use Ale and other tonics to improve the general health.

A Corset should be used made to fit the pelvis, to which is attached two bands, which pass between the thighs and are then fastened at both ends to the corset; having a pad in the centre for the support of the prolapsed organ.

Use the treatment given for Dyspepsia; it will tone up the system; there is nothing better. Also the *Salicylic Acid Mixture*. P. 272.

Or, the Arseniate of Gold [5th,] or, Alstonia Constricta 2d dilution.

Use the Stillingia Liniment.

392. "WEED" OR INFLAMED BREAST.

Keep the breast as free of milk as possible. Then in a glass of water put half a grain of Tartar Emetic, to which add ten drops of Aconite.

DOSE: a teaspoonful every hour or two until better; then extend the time. At the same time apply over the inflamed breast the Clay mixture given on P. 237.

Or, cut open a chicken and apply it warm; or a cat skin the same; there is nothing better.

394. CHILD-BED FEVER. [Puerperal Fever.]

Give [Acon. Bell., and Bry.]

Also take half a dram of Cream of Tartar, and the fourth of a grain of Podophyllin; mix, and give at a dose; to be repeated every two hours until it produces free purging.

To be followed by No. 1, P. 272.

Send for a good Physician.

395. WOUNDS, CUTS, ETC.

Cleanse the wound; remove foreign bodies; bring the injured parts together; hold them in place by adhesive plaster or stitches; keep quiet. Cover the wound with a doubly folded piece of lint soaked in Carbolic Acid Ointment, or Calendula Lotion. Do not disturb the wound too often; but keep it moist with the Ointment or Lotion. If the wound is extensive, or the blood comes in *spurts* and is *bright red*, send for a Surgeon at once; an artery is cut.

Until the arrival of the Surgeon, tie a handkerchief with a knot tied in it; around the limb near the wound (between it and the heart;) which place over the artery, [which may be found by the pulsation;] With a stick inserted under the handkerchief twist it tight. Should fever supervene, treat as fever. A few drops of Arnica or Hamamelis given occasionaly will be found useful. After suppuration Heper Sulph., Silicea, Calc c., or Sulph.

In case of Broken Bones, Dislocations, etc,, send for a Surgeon.

396. SPRAINS.

Bathe the parts in *hot* or cold Water, and apply Stramonium; or Arnica and Rhus tox. lotion.

Or, Carbolic Acid diluted in water.

Or, Aconite Liniment; keep the joint as quiet as possible.

397. BURNS AND SCALDS.

The success in treating burns and scalds depends on *excluding* the air from the burned part as soon as possible; and not removing the dressing until the cure is completed.

Saturate lint with *Carbolized Oil* [one of the first to ten of the latter] and cover the part : or, with the Carbolic ointment.

Or, dampen Soda or Soap and apply : or wet the part and cover with Starch or Flour ; or the white of an Egg, Mucillage, etc. If there is nervousness, pain or fever, treat as such ; as given under the appropriate head.

398. ELECTRICITY.

The *general principles* on which electricity is used for the cure of disease may be given briefly as follows :

Over INFLAMED or HEATED parts the POSITIVE POLE should be applied ; while the NEGATIVE pole is applied to the cold or inactive parts; pass the current down towards the chest ; if the parts are below, up toward the heart. Remember that the right hand or side of most persons is positive and the left negative. These few rules given, use your common sense in each separate case.

399. LIMITS OF MEDICINE.

Alas ! no herb in any garden grows
That can avert grim Death's unerring throes.
Were doctors skilled enough to undermine
Each fell disease, they'd almost be divine.
But, as all practice shows, no doctor can
Make life anew, though he may stretch its span.
Nature this power most jealously reserves,
Alone the body heals and life preserves,

APPENDIX.

CALLS UPON THE SICK.

"The following hints should be remembered by those calling upon the sick: only call at the door, unless you are sure your friend is able to see you without harm. Enter and leave the house, and move about the room, quietly. Carry a cheerful face, speak cheerful words. In order to cheer you need tell no lies. If your friend is very sick, do not fall into gay and careless talk in the attempt to be cheerful. Don't ask questions, and oblige your friend to talk. Talk about something outside, and *not about the disease and circumstances of the patient*. Tell the news, BUT NOT THE LIST OF THE SICK AND DYING. If possible, carry something with you to please the eye and remove the monotony of the sick-room; a flower, or even a picture which you can loan for a few days. If desirable, some little delicacy to tempt the appetite will be well bestowed. The perfumes of some flowers are poisonous, and they should never be carried into the sick-room. Especially is this true of the tuberose, oleander, heliotrope, hyacinth, orange, lilac, syranga, and lilies. Stay only a moment, or a few minutes at the longest, unless you can be of some help."

TABLE OF PROPORTIONATE DOSES FOR DIFFERENT AGES.

Under ½ year, 1-15 of full dose.   Under 7 years, ⅓ of full dose.
"   1   "   1-12   "   "   14   "   ½   "
"   2   "   ⅛   "   "   20   "   ⅔   "
"   3   "   ⅙   "   "   21   "   full dose
"   4   "   ¼   "   "   60   "   ¾   "

This table is designed for the remedies of the more heroical treatment.

POISONING AND HOW TO TREAT IT.

To save time in an emergency, the following table may be consulted, which gives the name of each of the most common poisons and the proper treatment for it:

Poison.	Treatment.
1. Unknown	Provoke repeated vomiting. Give bland liquids; Stimulate, if necessary.
2. Acids— Sulphuric, Nitric, Muriatic, Oxalic.	Give an Alkali; Provoke vomiting; Give bland fluids; Secure rest; Stimulate, if necessary.
3. Alkalies— Hartshorn, Soda, Potash, Lye	Give an acid (Vinegar); Provoke vomiting; Give bland liquids; Secure rest; Stimulate if necessary.
4. Arsenic— Paris green, Scheel's Green,	Provoke vomiting;* Give dialyzed Iron and Salt;* Give a dose of Castor Oil; Secure rest; Stimulate if necessary.
5. Sugar of Lead	Give Epsom Salts;* Provoke vomiting;* Give bland liquids; Give a dose of Castor Oil.
6. Corrosive Sublimate, Tartar Emetic.	Provoke vomiting;* Give Strong Tea, without milk;* Give raw Eggs and Milk; Give a dose of Castor Oil; Stimulate, if necessary.
7. Phosphorus	Provoke vomiting; Give five grain doses of Sulphate of Copper; Give dose of Magnesia, but no Oil.

8. Opium— 　Laudanum, 　Paregoric, 　Chloral.	Provoke vomiting, repeatedly; Give Strong Coffee without Milk; Keep up the breathing.
9. Strychnia..................	Provoke vomiting, once or twice; Give a purgative; Secure absolute quiet.
10. Aconite..................	Provoke vomiting; Stimulate well;
11. Lunar Caustic (Nitrate of Silver.)	Give strong Salt and Water;† Provoke vomiting;†
12. Alcohol...................	Provoke vomiting; Give Hartshorn and Water
13. Jamestown Weed, 　Hemlock, 　Nightshade (Belladona,) 　Toadstools, 　Tobacco.	Provoke vomiting; Stimulate well.
14. Decayed Meat and Vegetables.	Provoke vomiting; Give a purgative; Give powdered Charcoal.

To promote vomiting, warm Water may be used, with a tablespoonful each of Mustard and Salt to the pint of water; give large quantities, and thrust the finger down the throat. *Bland liquids* are raw Eggs, Milk, Oils, Elm and Gruel. *Stimulants* are Tea, Coffee, Whiskey, Brandy, Ammonia, Cajeput, etc. *Alkalies* as *Antidotes* are Ammonia, Soap, Lime, Soda, Chalk, Plaster, Magnesia, Powdered Charcoal and Wood Ashes. *Acids* as *Antidotes* are Vinegar and Lemon juice. Success depends on giving the antidote *quickly* before the poison can act upon the system.

---

* Repeat several times.　　† Repeat many times.

DIRECT METHOD FOR RESTORING THE APPARENTLY DROWNED.

1st. INSTANTLY turn the patient downwards with a large firm roll of clothing under the stomach and chest.

Press with your weight two or three times, for four or five seconds each time, upon the patient's back, so that the water is pressed out of the lungs and stomach, and drains freely downwards out of the mouth. Then

2d. QUICKLY turn the patient face upwards, the roll of clothing put under his back just below the shoulder blades, the head hanging back as low as possible.

Place the Patient's hands together above his head.

Kneel with patient's hips between your knees.

Fix your elbows against your hips.

Now, grasping the lower part of the patient's chest, squeeze the two sides together, pressing gradually forward with all your weight, for about three seconds, until your mouth is nearly over the mouth of the patient; then, with a push, *suddenly* jerk yourself back—

Rest about three seconds; then, begin again.

Repeat these bellows-blowing movements, so that air may be drawn into the lungs, about eight or ten times a minute.

Remember, the above directions must be USED ON THE SPOT, the instant the patient is taken from the water. A moment's delay—and success may be hopeless. As soon as the water is pressed from the lungs, all clothing should be ripped away from the chest and throat. In making the pressure either for the removal of water or for breathing, increase it *gradullay* and *suddenly* let go with a jerk. With women and children use less force.

DO NOT STOP THESE MOVEMENTS UNDER AN HOUR, unless the patient breathes. Be careful not to interrupt the first short natural breaths. If they be long apart, carefully continue between them the bellows-blowing movments as before.

After breathing is regular, keep the patient warm with blankets, rubbing with warm hands, etc.

Prevent crowding around the patient; plenty of fresh air is all-important.

Spirits and water only, in occasionally small doses may now be given; if hot the better. After this encourage quiet and sleep. [Dr. Benj. Howard.]

"The apparently drowned" have been restored after *many* hours of apparent death; by constant unwearied efforts, as above given.

THE STOMACH CLOGGED with *nut* kernels; to cure; eat freely of Salt.

SYRUPS AND BALSAMS.

1. LUNG BALSAM.—Sunflower Seed 2½ ℔, Marshmallows 1℔, Sanguinaria, 1 oz., Peach Kernels 4 oz., Hydrastis 2 oz; all to be pulverised and put into 2 qts. of the best gin; keep it in a closed vessel for two weeks; strain, and add cold water to the residue, and strain again until ½ a gallon has passed, which set aside; add water and repeatedly boil and strain until the strength is extracted, and there is 2 gallons of the liquid: add 12℔ of white Sugar, and carefully evaporate until but two gallons remains; now add Gum Arabic 1℔, and Sulphate of Morphine one dram, (rub-

bed up in Syrup,) and then add the first liquid which was set aside.

DOSE : From a teaspoonful to two tablespoonsful, three times a day, or as often as the case demands. There is nothing superior.

2. CROUP SYRUP.—Syrup of Squills 3 oz., Syrup of Lobelia 2 oz., Syrup of Ipecac 1½ oz., Hoffman's Anodyne 2 drams, Oils of Cinnamon, and Anise 3 drops each.

DOSE : one teaspoonful every five minutes with warm water, until *free vomiting* ensues ; repeat as often as necessary. This is invaluable in cases of croup.

OINTMENTS AND SALVES.

1. CARBOLIC ACID OINTMENT.—Carbolic Acid 25 drops, Eye-Salve (No. 4.) six drams. Mix. This is an efficient ointment in small-pox, ulcers, and all skin diseases.

2. CUTIS OINTMENT.—Chrysophanic Acid one dram, Lard or Vaseline one ounce. Put them together and place the vessel in boiling water for half an hour, stir constantly. When "set," mix with pestle and mortar. For Psoriasis and other skin diseases.

3. ALL HEALING OINTMENT.—Take Stramonium leaves half a peck, wilt, crush and stew them in half a pound of Lard ; strain and add gum Turpentine, Honey, and Bees' Wax, five ounces of each ; melt and stir in four ounces of finely pulverised Verdigris.

For *general* use this is *very* good. Vaseline or Cosmoline may be substituted for the Lard or cerates in any of the above.

4. HAWK EYE SALVE.—White Oxide of Zinc half an ounce, Sulphate of Morphine twenty-five grains, Benzoic Acid fifteen grains. Mix by *long trituration* in a mortar. Now take Spermaceti one and a half ounces, white Wax half an ounce, Olive Oil four ounces; melt, and add the first mixture, and apply heat until it boils; remove from the fire and constantly stir until cold.

For most kinds of sore eyes this can hardly be surpassed. It is valuable in piles, burns, ulcers, etc.

To be used on the lids of the eyes three times a day, especially at night.

5. MAGIC; SNUFF; OR FEVER OINTMENT.—Snuff or Tobacco, Lobelia Seeds, Raisins and Lard equal weights. Simmer together, strain and press out while hot.

For Croup, inflammation of the lungs, bowels, etc.

This is magic in its effects.

PILLS AND POWDERS.

1. GENERAL CATHARTIC PILLS.—Podophyllin 1 dram, Colocynth 2 drams, Aloes 1½ ozs., Gamboge 6 drams, Castile Soap 3 drams, Oils of Cinnamon and Pepermint 6 drops each. Mix, and make into five grain pills.

Dose: from one to four. Experience, has proven this to be a valuable pill.

2. DYSENTERY PILL.—Leptandrin, Rhubarb, Ipecac and Castile Soap 2 drams each; Opium 1 dram. Mix, and make into three-grain pills.

Dose: one every hour until relieved. This is almost a specific in Dysentery.

3. SWEATING PILLS.—The soft extract of Quaken Aspen Bark 2 drams, Lobelia Seed (Pulv.) 1 dram, Cayenne Pepper ½ dram. Thoroughly mix by long pounding in a mortar, and make into two grain pills.

Dose: one every half hour until the degree of perspiration required is had. Warm tea increases their action. These are woderfully effective diaphoretic pills.

4. LIVER PILLS.—Podophyllin 10 grains, Leptandrin 20 grains, Hydrastin, Capsicum, Mercurius v., and Salicylic Acid 20 grains each. Mix, and make into 60 pills.

Dose: from one to three every two hours as a physic; in chronic disease, one at night.

5. ANTIBILIOUS POWDERS.—Alexandria Senna, and Jalap 1 ℔ each; Ginger 4 ozs., Pulverise and mix.

Dose:—a teaspoonful in a little hot water, sweetened and taken when cold; repeat the dose if necessary. A mild efficient cathartic; equaled by few, and surpassed by none.

6. SWEATING ANODYNE POWDER.—Take of powdered Opium and Ipecac 15 grains each, Camphor 30 grains, Super Carbonate of Soda, or Cream of Tartar ¼ oz. Mix thoroughly.

Dose:—from 5 to 20 grains for an adult, in proportion for children, repeat in from one to four hours, according to the urgency of the case. It is an excellent anodyne, relieving pain, causing moderate perspiration and a disposition to sleep. It is preferable to the common Dover's Powder, and it will be found acceptable when other preparations of Opium have failed.

7. ALTERATIVE POWDERS.—Sanguinarin, Podophyllin, Leptandrin, Caulophyllin and Baptisin 1 grain each; China (3,) and Mercurius v. (3,) ½ oz. each; thoroughly incorporate by long trituration in a mortar.

DOSE: What will lie on a ten cent piece, morning and night.

8. RESTING POWDERS AND DROPS.—Atropia (3,) or Cina (3.)

9. ALSTONIA CONSTRICTA—CHINOIDINE AND EUCALYPTUS.—All have gained favor as substitutes for Quinine.

10. AGUE PILL.—Add to the Liver Pill No. 4., 1 dram of Quinine and sufficient Oil of Pepper to mix. Given as other ague medicines.

1. STILLINGIA — CAJEPUT — ANTISPASMODIC LINIMENT.—Oils of Stillingia and Cajeput 1 oz. each; Oil of Lobelia 2 drams; Tinc. of Camphor 2 ozs. Mix together. This is a *powerful* liniment: is used to great advantage in *Asthma*, Croup, Spasmodic diseases, Rheumatism, etc.

2. STIMULATING LINIMENT.—Tinc. of Capsicum 2 ozs., Tinc. of Opium. Aqua Ammonia, Oils Origanum, and Cajeput each 2 drams; Oil Cinnamon 1 dram, Tinc. Camphor, Hoffman's Anodyne ½ oz. each. Mix. A powerful Anodyne and Stimulating Liniment. Used in paralysis and like affections, stimulating the nervous system.

3. BATHING DROPS.—Tinc. of Hamamelis, Arnica and Stimulating Liniment equal parts. Sprains, Bruises, internal inflammation, etc.

4. CHEAP LINIMENT.—By adding one ounce of any of the above to ten ozs. of any of the following, you have a good cheap Liniment for general use. Crude Coal Oil, Olive Oil, a weak solution of Carbolic Acid, Turpentine, Alcohol, Tinc. Calendula, Arnica, etc.

TONICS AND OTHER FORMULAS.

1. WILD CHERRY BITTERS.—Wild Cherry Bark, Virginia Snake Root, Valerian Root, and Gentian, ¼ oz. each; steep and use as a bitters: a tablespoonful when needed. Reliable.

2. GENERAL TONIC.—Hydrastis c., Gentian and Valerian 1 oz. each, repeatedly boil until the strength is extracted; strain and make a quart of Syrup with Sugar; now add ten drops of the tincture of Nux Vomica, and one dram of Carbonate of Ammonia.

DOSE: from a teaspoonful to a tablespoonful three times a day. Gives tone to the digestive organs. Choice.

5. DIARRHEA AND DYSENTERY MIXTURE.—Take 1 oz. each of Tinc. Ipecac, Laudanum, Spirits of Camphor, Hoffman's Anodyne, Tinc. Ginger, Spirits of Nitre, Tinc. Rhubarb, Ess. of Peppermint; and 4 oz. of Brandy. Mix.

DOSE: one to two teaspoonsful in Syrup.

3. TONIC ALTERATIVE. See P. 244. Also Baptisin, Leptandrin, Podophyllin, Quinine, Macrotine and Caulophyllin one grain each, rubbed up with ½ oz. of Sugar of Milk. Used in female debility, etc. Very good.

DOSE: what will lie on a dime.

4. GENERAL ALTERATIVE.—To equal parts of Compound Syrup of Stillingia (See Dispensatory,) and the second mixture on P. 244, add Iodide of Potassium two drams to the pint. This is one of the very *best* of Alteratives.

DOSE: from a teaspoonful to a wine-glassful; from once to three times a day.

THE LANGUAGE OF FLOWERS.

"Language of Flowers" will be the means of *improving* the *health* of girls, by inducing them to engage in their culture.

Amaranth—Immortality.
Aster—Beauty in retirement.
Ash—Grandeur.
Alyssum—Worth beyond beauty.
Bachelor's Button—Hope in misery.
Balm—Sweets of social intercourse.
Balm of Gillead—I am cured.
Balsam—Impatience.
Bay Leaf—I change but in dying.
Birch—Gracefulness.
Bindweed—Humility.
Blue Bell—Constancy.
Broome—Neatness.
Burdock—Importunity.
Calla—Feminine modesty.
Chamomile—Energy in adversity.
Cardinal Flower—Distinction.
Carnation—Pride.
Catchfly—A snare.
Cinquefoil--Love, constant but
Hydrangea—Heartlessness.
Ice Plant—Your looks freeze me.
Ivy—I have found one true heart.
Jasmine—Amiability.
Jonquil—Affection returned.
King-cup- I wish I was rich.
Lady's Slipper—Capricious beauty
Laurel—Virtue is true beauty.
Lavender—Acknowledgment.
Lemon--Discretion.
Lilac—First emotion of love.
Lily—Purity.
Lily of the valley—Heart withering in secret.
Locust--Affection beyond the grave
Mallows--Sweet disposition.
Maple—Reserve.
Marigold—Contempt.
Mignonette—Moral beauty.
Moss—Maternal love.

APPENDIX.

hopeless.
Clematis—Mental excellence·
Columbine—I cannot give thee up.
Corn—Riches.
Coreopsis—Always cheerful.
Coriander—Concealed merit
Cypress—Disappointed hopes.
Dahlia—Elegance and dignity.
Daisy—Beauty and innocence.
Dandelion—Coquetry.
Elder—Compassion.
Evergreen—Poverty and worth.
Fir—Time.
Forget-me-not—True love.
Fuchsia—Humble love.
Geranium, Rose—Preference
" Scar.—Thou art chang'd
" Oak—True friendship.
" Lemon—Tranquility of mind.
Gilly Flower—Lasting beauty
Golden Rod   Encouragement.
Grape—Charity.
Hawthorn—Hope.
Hazel—Reconciliatio -
Heliotrope—Devotion.
Hibiscus—Beauty in vain.
Hollyhock—Ambition.
Honeysuckle—Fidelity.
Hop—Injustice.

Nasturtium---Patriotism.
Nightshade---Dark thonghts.
Pea, Everlasting---Wilt thou go?
" Sweet---Departure.
Peach Blossom-- I am your captive.
Phlox---Our souls are united.
Pine---Time and faith.
Pink, White---Lovely and pure affection.
" Red---Woman's love.
Polyanthus --Confidence.
Primrose---Modest worth.
Rose-bud---Confession of love.
" Bridal---Happy love.
" Damask---Bashful love.
" Multiflora---Grace.
" White---Too young to love.
" Red-leaved---Diffidence.
Sage---Domestic virtues.
Sunflower---Smile on me still.
Sweet William---Gallantry.
Thistle---Never forget.
Tulip---Beautiful eyes.
Verbena---Sensibility.
Violet---Faithfulness.
Wallflower---Fidelity in misfortune
Willow---Forsaken.
Woodbine -Fraternal love.
Yarrow-- A cure for the heartache.
Zinnia---I mourn your absence.

A TABLE OF WEIGHT-MEASURES.

The following table gives the approximate weight-measure of medicines.

A teaspoonful is about one fluid dram, or sixty drops.
A dessertspoonful is about two fluid drams.
A tablespoonful is nearly equal to five fluid drams.
A wine glass holds about four fluid ounces.
A pint is very nearly equal to a fluid pound.

APPENDIX.

HERSCHEL'S WEATHER TABLE

FOR FORETELLING THE WEATHER, THROUGHOUT ALL THE LUNATIONS OF EACH YEAR, FOREVER.

By the use of the "weather table and the water-proof clothing" sickness may often be prevented by avoiding unnecessary exposure.

This table and the accompanying remarks are the result of many years' actual observation, the whole being constructed on a due consideration of the attraction of the Sun and Moon, in their several positions respecting the Earth, and will, by simple inspection, show the observer what kind of weather will most probably follow the entrance of the Moon into any of its quarters, and that so near the truth as to be seldom or never found to fail.

If the New Moon, First Quarter, Full Moon, or Last Quarter, happens	IN SUMMER.	IN WINTER.
Between midnight & 2	Fair....	Frost unless wind S-West.
" 2 & 4 morning	Cold and showers...	Snow and stormy.
" 4 & 6 "	Rain................	Rain.
" 6 & 8 "	Wind and rain......	Stormy.
" 8 & 10 "	Changeable.........	Rain if wind W. snow if E.
" 10 & 12 "	Frequent showers....	Cold and high wind.
" 12 & 2 aftern'n	Very rainy..........	Snow or rain.
" 2 & 4 "	Changeable.........	Fair and mild.
" 4 & 6 "	Fair................	Fair.
" 6 & 8 "	Fair if wind North-W	Fair, Frosty if wind N. N-E
" 8 & 10 "	Rainy if South or S-W	Rain or snow if S. or S-W
" 10 & midnight	Fair................	Fair and frosty.

OBSERVATIONS.—1. The nearer the time of the Moon's change, first quarter, full and last quarter are to *midnight* the fairer will be the weather during the next seven days.

2. The space for this calculation occupies from ten at night till two next morning.

3. The nearer to *midday* or *noon* the phases of the moon happens, the more foul or wet weather may be expected during the next seven days

4. The space for this calculation occupies from ten in the forenoon to two in the afternoon.

These observations refer principally to the Summer, though they affect Spring and Autumn nearly in the same ratio.

5. The moon's change, first quarter, full and last quarter, happening during six of the afternoon hours, i. e., from four to ten, may be followed by fair weather ; but this is mostly dependent on the *wind*, as is noted in the table.

6. Though the weather, from a variety of irregular causes, is more uncertain in the latter part of Autumn, the whole of Winter, and the beginning of Spring, yet in the main the above observations will apply to those periods also.

7. To prognosticate correctly, especially in those cases where the *wind* is concerned, the observer should be within sight of a good *vane*, where the four cardinal points of the heavens are correctly placed. A rosy sky at sunset, whether clear or cloudy, indicates *fine weather*. An Indian-red tint at sunset foreshadows *rain*. A red sky in the morning bad weather, or much wind—perhaps rain. A gray sky in the morning, fine weather.

WATER-PROOF CLOTHING.

It is very desirable to understand how to make overcoats, horse blankets, hay caps, grain bags, and anything else water-proof, by incurring a little expense. We herewith give a recipe which has often been sold for hundreds of dollars to unsophisticated clothiers : To about ten quarts of rain water add a half pound of powdered alum ; stir this at intervals until it becomes clear ; then pour it off into another bucket and put the garment therein ; let it remain for twenty-four hours, then hang it out to dry without wringing it.

CONSUMPTION PILL.

Evaporate *Crude* Coal-Oil to the consistency of **Wax**, and make into pills with **Hydrangea**. Take one, four times a day, every other day ; it often cures consumption.

LIST OF HOMŒOPATHIC REMEDIES ;

Their potency and abbreviated name. ‡*Most* in use ; † not used so often; while those unmarked are none the less valuable ; but not so often required. The figures represent the potency of the medicines ; next abbreviated name ; and then the full name.

1	Acon.....Aconitum‡		5	Graph....Graphites †
2	Alst. con...─.Alstonia con†		Tinc.	Ham....Hamamelis v.‡
2	Apis.....Apis mellifica†		5	Hep. Sulph....HeparSulph‡
2	Apocy.....Apocynum can.		Tinc.	Hydras....Hydrastis.†
Tinc.	Arn.....Arnica mont.‡		2	Hyoscy....Hyoscyamus n.†
5	Arsen. g....Arseniate Gold		2	Igna....Ignatia a.‡
4	Ars.....Arsenicum alb.‡		2	Ipec....Ipecacuanha.‡
3	Atrop.....Atropia Sulph.‡		1	Jalap....Jalappa.
5	Arum....Aurum met †		2	Kali bic...Kali bichrom.
Tinc.	Bapt....Baptisia tinct‡		4	Lach....Lachesis.†
4	Baryt....Baryta carb.		1	Lept.....Leptandrin.‡
2	Bell.....Belladonna.‡		1	Lobe....Lobelia inf.‡
1	Brom. c..Bromohydrate caff.		4	Lycopo....Lycopodium.†
2	Bry....Bryonia alba‡		1	Mac....Macrotine.‡
3	Cact.....Cactus grand.†		3	Merc. v....Mercurius vivus‡
3	Calc.....Calcarea carb.‡		5	Merc. c.....Mercurius cor.†
Tinc.	Calend....Calendula off‡		3	Mer. Iod..Mercurius Iodium.
Tinc.	Camph....Camphora.‡		3	Mezer....Mezereum.
3	Canch....Canchalagua. •		6	Natr. m....Natrum mur.†
3	Cann....Cannabis Sativa.†		4	Nitric a....Nitric acid†
2	Canth....Cantharis.‡		3	Nux v. Nux vomica‡
4	Carbo....Carbo veget.†		2	Opi....Opium.†
1	Caul.....Caulophyllin.‡		5	Petrol. Petroleum.‡
3	Caust....Causticum.†		3	Phos....Phosphorus‡
Pure	Ceri. ox.....Cerium oxide.		3	Phyto....Phytolacin-†
2	Cham....Chamomilla.‡		2	Plant....Plantago major.
3	China....China off.‡		1	Pod...Podophyllin.‡
2	Cina....Cina.‡		4	Puls....Pulsatilla.‡
3	Coccu....Cocculus.†		3	Rhus tox.Rhu stoxicodendr'n
2	Colch....Colchicum.†		3	Sabin....Sabina.
2	Coloc....Colocynthia†		3	Sangui....Sanguinarin.‡
Tinc.	Corall...Corallorhiza.		1	Santon....Santonium.†
4	Cupr. m....Cuprum met.†		1	Sarsap....Sarsaparilla.
3	Crocus s....Crocus Sativa.†		2	Senec....Senecin.
1	Diosc....Dioscorin·†		4	Sepia....Sepia.†
2	Dulc....Dulcamara.†		4	Silic....Silicca.‡
3	Ergot....Ergotin.		2	Spong. t....Spongia tosta.‡
2	Euph....Euphrasia		4	Sulph....Sulphur.‡
4	Ferri....Ferrum m.†		4	Thuja....Thuja occid.
Tinc.	Gels.....Gelseminum.‡		3	Tart. emet..Tartar emetic.‡
5	Glon....Glonoinum.†		3	Verat. a..Veratrum album.†

TELL YOUR WIFE.

"Women are more keen than men to see approaching evil, and far more heroic in bearing it.

If you are in any trouble or quandery, tell your wife—that is if you have one—all about it at once. Ten to one her invention will solve your difficulty sooner than all your logic. The wit of woman has been praised, but her instincts are quicker and keener than her reason. Counsel with your wife, or your mother or sister, and be assured that light will flash upon your darkness. Women are too commonly adjudged as verdant in all but purely womanish affairs. No philosophical students of the sex thus judge them. Their intuitions, or insights, are the most subtle, and if they cannot see a cat in the meal, there is no cat there. In counseling a man to tell his wife, we would go farther, and advise him to keep none of his affairs a secret from her. Many a home has been happily saved, and many a fortune retrieved, by a man's full confidenc'in his "better half." Woman is far more a seer and prophet than man, if she be given a fair chance. As a general rule, wives confide the minutest of their plans and thoughts to their husbands, having no involvments to screen from them. Why not reciprocate, if but for the pleasure of meeting confidence with confidence? We are certain that no man succeeds so well in the world as he who, taking a partner for life, makes her the partner of his purposes and hopes. What is wrong of his impulse or judgment she will check and set right with her almost universally right instincts. "Help-meet" was no insignificant title as applied to man's companion. She is a help-meet to him in every darkness, difficulty and sorrow of life. And what she most craves and most deserves is confidence—without which love is never free from a shadow."

INDEX.

Abscess, 82.
Abortion, 23.
Agues, 24, 175.
Alkalies, and Acids 224.
Alteratives, 297, 299.
Apoplexy, 29, 210.
Appendix, 289.
Appetite Canine, 30
Asthma, 30, 249.
Baldness, 33.
Barber's Itch, 233.
Bathing, etc , 151, 138.
Balsam Lung, 293.
Balsam Turlingt'ns 135
Boils, 35.
Bleeding of Nose 33, 220
Blood Spitting, 34, 245.
Blisters, 35, 235.
Blues, 253.
Breasts hard &c 35, 286
Bruises, 36, 215.
Burns & Scalds, 36, 287.
Breath offensive, 222.
Bronchitis, 238, 51.
Calls upon the Sick 289.
Cancers, 37, 274.
Catarrh, 218.
Carbuncle, 227.
Chaps in Nipples, 42.
Chaped Hands, 43, 235.
Chafing of Infants, 188.
Chilblains, &c. 40, 234.
Children Dis. of, 40, 179
Clergyman's Throat 238
Cholera and Cholera Morbus, 42, 258.
Cholera Infantum, 197.
Chicken-Pox, 184.
Colic, 44, 189, 257.
Colds, 47, 43, 238.
Cold Feet, 282.
Consumption, 47, 242.
" Pills, 302.
Convulsions 49, 195.
Corns, 50, 235.

Costiveness, 50, 264.
Coughs, 51, 293.
Cramp, 53.
Crying, Colic, of Infants, 189, 44.
Croup, 185.
Croup Syrup, 294.
Cuts, &c., 54, 286.
Deafness, 54, 218.
Delivery, 56.
Diabetes, 56.
Diarrhea, 196, 263.
Diphtheria, 192.
Doses, 266, 289.
Dropsy, 56, 273.
Drowned, 59, 292.
Dysentery, 199, 264.
Dyspepsia, 252.
Ear ache, &c , 59, 217.
Electrifying, 139, 288.
Eyes, diseases of, 60, 192, 213, 215, 216.
Eye Salve and Water, 137, 295, 64.
Erysipelas, 27, 207.
Fainting, 66.
Falling Sickness, 66, 212
Falling of Fundament, 67.
Falling of Womb, 67. 285.
Fat, 68.
Felon, &c., 130, 236.
Fever, (General,) 156. 68.
" Cause. 157.
" Symptoms, 160.
" General Treatment, 167.
" Continued 172
" Bilious, 172.
" Typhoid, 173.
" Typhus, 175.
" Intermittent, 175

" Scarlet, 180.
" Yellow, 201.
" Lung, 241.
" Rheumatic, 272.
" Child bed, 286.
Fistula, 70, 232.
Fleas, &c., 71.
Flooding, in lying-in, 72, 284.
Flux, 73, 199, 42.
Food, 149.
Gangrene, 74.
Gout, 75, 273.
Goitre, 226.
Gravel, &c., 76, 268.
Green Sickness, 76, 283
Gums, 77, 223.
Hair to Grow, 77.
Hawk Eye Salve, &c., 295, 64.
Head-ache, 77, 209.
" Stoppage in, 79.
Heart-burn, 79, 252.
Hiccough, 80.
Hives, 184.
Hoarseness, 80, 185 238.
Homœopathic Med. List, 303.
Homœopathic Dose, &c 266.
Hysteric Disorders, 81.
Iliac Passion, 82.
Influenza &c., 237.
Inflammation, 205.
" Symptoms, 206.
" of Brain, 208.
" of Lungs, 241.
" of Liver, 262.
" of Kidneys 267.
" Bladder, 267.
" Spleen, 271.
" Breasts, 286.
Itch, 82, 233.
Jaundice, 81, 261.

INDEX.

King's Evil, 83, 274.
Lameness, 84.
Laws of Health, 19, 148.
Language of Flowers. 299.
Legs, Sore, 84.
Leprosy, 85.
Lethargy, 86.
Lice, to kill, 86,
Lightning, Apparent Death by, 86,
Limits of medicine 288
Liniments, &c., 297.
Liver Complaint, 262
Liver Pad, 177.
Lues Venera, 86, 274.
Lunacy, 87.
Lung Balsam, 293.
Madness, 87.
Mad-Dog, bite, 88, 277.
Measles, 89, 182.
Menses, deranged, etc. 89, 282.
Medicines, Purity of 147, 134.,
Medicines, using, 17, 266.
Milk, 91.
Mortification, 91.
Morning Sickness, 256.
Mumps, 183.
Nervous Disorders, 91, 278, 141.
Nettle Rash, 92, 184.
Night sweats, 244, 117.
Neuralgia, 281.
Nurse, 147.
Old Age, 93.
Offensive breath, 222.
Ointments, &c., 294.
Pains, 93, 115, 271.
Palsy, 93.
Palp. of Heart, 95, 247.
Piles. 95, 270.
Pills and Powders, 295.
Plague, 96.

Pleurisy, 97, 241.
Poisoning, to treat, 97, 290.
Polypus, 98.
Preface and Introduction, 5, 9, 145.
Prick or Cut, 98.
Ptyalism, 98.
Purges, 99, 295, 296.
Phlegm, 72.
Quinsy, 99, 226.
Rheumatism, 100, 272.
Rickets, 102.
Ring Worms, 102, 236.
Running at the Nose. 102.
Rupture, 102, 271.
Scald Head, 104, 212
Scotch Pills, 135.
Scrotum Swelled, 105.
Scrofula, 83, 274.
Scurvy, 106, 223, 105.
Shin, broken, 107.
Shingles, 107, 231.
Sickishness, 108,
Sinews Shrunk, 108.
Skin Rubbed off, 108,
Skin Diseases, 230.
Sleeping, 149.
Sleeplessness, 123, 189, 297.
Slow Growth, 192.
Small-Pox, 108, 203.
Snake-bite, 123, 278.
Snuffles of Infants 191.
Sore Throat. 110, 238 192, 226.
Sores, 109, 105,.
Sprains, 111, 287.
Stings, 111, 278.
Stitch in the Side, 113.
Stomach, 113, 252.
Stone, 114, 268.
Stoppage in the Kidneys, 116, 268.
Stranguary, 116, 268.
Sunburn, 116.

Sunstroke, 212.
Surfeit, 116.
Sweating, 116.
Swellings, 117.
Sciatica, 104.
Table of Measures, 300
" of Doses. 289, 266
" of Poisons, 290.
" of Flowers. 299.
" of Weather, 301.
" of Homeop. Med. 303.
Tape Worm, 269, 131.
Teeth, 118, 194.
Tetter, &c., &c., 230.
Tell Your Wife, 304.
Testicles, pain in, 121.
Thirst, 120.
Thrush, 121. 190.
Thorns, &c., 121.
Tonsils Swelled, 122, 226.
Tooth Ache, 119, 221.
Tonics & Mixtures 298.
Topor of Limbs, 122.
Tympany, 122.
Ulcers, 124, 227.
Urine, 126, 268.
Uvula, 127.
Vein Cut, 122.
Venereal, 274.
Vertigo, 123.
Vigila, 123.
Vomits, &c. 126, 135, 256
Water-Proof cloth 302.
Warts, 128, 235.
Weather Table, 301.
Weakness in Ancles, 128
Weed, 286, 35.
Wen 128.
Wetting Bed, 191, 268.
Whooping Cough. 41, 187.
Whites, 129, 283.
Whitlow, 130, 236.
Worms, 130, 188.
Wounds, 34, 132, 286.

www.ingramcontent.com/pod-product-compliance
Lightning Source LLC
Chambersburg PA
CBHW030815230426
43667CB00008B/1227